A TEXT BOOK OF

SWITCHGEAR AND PROTECTION

(Elective-III)

FOR

SEMESTER – VI

**THIRD YEAR (T.Y.) B. TECH COURSE IN
ELECTRICAL ENGINEERING / ELECTRICAL ENGINEERING
(ELECTRONICS AND POWER) / ELECTRICAL AND ELECTRONICS
ENGINEERING / ELECTRICAL AND POWER ENGINEERING**

**Strictly According to New Revised Credit System Syllabus
of Babasaheb Ambedkar Technological University (BATU),
Lonere, (Dist. Raigad) Maharashtra,**
(w.e.f. June 2019-20)

L. KALYANKUMAR
M. Tech (Electrical Power Systems)
Assistant Professor,
Department of Electrical Engineering,
M.B.E.S. College of Engineering,
AMBAJOGAI.

N1114

SWITCHGEAR & PROTECTION (ELECT., BATU) ISBN : 978-93-89944-02-O

| First Edition | : | January 2020 |
| © | : | Author |

Published By :
NIRALI PRAKASHAN
Abhyudaya Pragati, 1312 Shivaji Nagar
Off J.M. Road, PUNE 411005
Tel : (020) 25512336/37/39
Email : niralipune@pragationline.com

➢ DISTRIBUTION CENTRES

PUNE

Nirali Prakashan (Local) : 119 Budhwar Peth, Jogeshwari Mandir Lane, Pune 411002, Maharashtra
Tel : (020) 2445 2044, Mobile : 9657703145, Email : niralilocal@pragationline.com

Nirali Prakashan (Outstation) : S. No. 28/27 Dhayari, Near Asian College, Dhayari, Pune 411041, Maharashtra
Tel : (020) 2469 0204, Fax : (020) 2469 0316, Mobile : 9657703143
Email : bookorder@pragationline.com

MUMBAI

Nirali Prakashan : 385 S.V.P. Road, Rasadhara Co-op. Hsg. Society Ltd., Girgaum, Mumbai 400004, Maharashtra
Tel : (022) 2385 6339 / 2386 9976, Fax : (022) 2386 9976, Mobile : 9320129587
Email : niralimumbai@pragationline.com

➢ DISTRIBUTION BRANCHES

JALGAON

Nirali Prakashan : 34 V. V. Golani Market, Navi Peth, Jalgaon 425001, Maharashtra
Tel : (0257) 222 0395, Mob : 94234 91860, Email : niralijalgaon@pragationline.com

KOLHAPUR

Nirali Prakashan : New Mahadvar Road, Kedar Plaza 1st Floor, Opp. IDBI Bank
Kolhapur 416012, Maharashtra. Mobile : 9850046155
Email : niralikolhapur@pragationline.com

NAGPUR

Nirali Prakashan : Above Maratha Mandir, Shop No 3, Second Floor,
Rani Jhanshi Square, Sitabuldi, Nagpur 440012, Maharashtra
Tel : (0712) 254 7129, Email : niralinagpur@pragationline.com

DELHI

Nirali Prakashan : 4593/15 Basement, Agarwal Lane, Ansari Road, Daryaganj
Near Times of India Building, New DelhiV 110002 Mobile : 8505972553
Email : niralidelhi@pragationline.com

BENGALURU

Nirali Prakashan : Maitri Ground Floor, Jaya Apartments, No. 99, 6th Cross, 6th Main,
Malleswaram, Bengaluru 560003, Karnataka
Mobile : 9449043034, Email : niralibangalore@pragationline.com

niralipune@pragationline.com | www.pragationline.com
Also find us on [f] www.facebook.com/niralibooks

Dedicated to...

My Students

....Author

PREFACE

It gives me great pleasure to present the book **"Switchgear and Protection" (Elective-III)** for the students of **Semester VI Third Year (T.Y.) B. Tech. Course Electrical Engineering / Electrical Engineering (Electronics and Power) / Electrical and Electronics Engineering / Electrical and Power Engineering of Dr. Babasaheb Ambedkar Technological University (BATU), Lonere, Dist. Raigad (Maharashtra)**. This book is strictly as per the new revised syllabus 2019-20 Pattern, effective from the Academic Year July 2019-20.

In New Revised Syllabus, there will be Internal Assessment (20 Marks), Mid Sem. Exam. (20 Marks) and End Sem. Exam. (60 Marks). End Sem. Exam. will be based on all six units and each unit will carry 12 Marks. The Theory Course will have 3 Credits.

The basic objective of this book is to bridge the gap between the vast contents of the reference books, written by the renowned International Author and the concise requirements of Undergraduate Students. This book has been written in a comprehensive manner using Simple and Lucid language, keeping in mind students' requirements. The main emphasis has been given on exploring the basic concepts rather than merely the Information. Solved Examples and Exercises have been provided throughout the book and at the end of the Unit. Also, I have given **Model Question Papers** for practice at the end of book.

I am also thankful to **Dr. B. I. Khadakbhavi** (Principal, MBES COE, Ambajogai) and **L. V. Bagale** (HOD, Electrical Engg. Deptt.), for their motivational support. Support and wishes of my **Family Members, Friends** have gone a long way in making this book possible.

I take this opportunity to express my thanks to Shri. Dineshbhai Furia, Shri. Jignesh Furia, Mrs. Nirali Verma, Shri. M. P. Munde and entire team of Nirali Prakashan, namely Mrs. Deepali Lachake (Co-ordinator), and her colleagues who really have taken keen interest and untiring efforts in publishing this text.

Suggestions of my esteemed readers are most welcome and will be highly appreciated.

Pune   **Author**

SYLLABUS

Unit I : Switchgear and Protection   8 Hrs.

Different types of switchgear, modes of classification, ratings and specifications. Protective Relaying: Need of protective relaying in power system, General idea about protective zone, Primary and backup protection, Desirable qualities of protective relaying, Classification of relays, Principle of working and characteristics of attracted armature, balanced beam, induction, disc and cup type relays, induction relays, Setting characteristics of over current; directional, differential, percentage differential and distance (impedance, reactance, mho) relays, introduction to static relays, advantages & disadvantages.

Unit II : Circuit Interruption   6 Hrs.

Principles of circuit interruption, arc phenomenon, A.C. and D. C. circuit breaker, Restricting and recovery voltage. Arc quenching methods. Capacitive, inductive current breaking, resistance switching, Auto reclosing Circuit Breakers: Construction, working and application of Air blast, Bulk oil, Minimum oil, SF6 and vacuum circuit breakers, Circuit breaker ratings, Rewritable and H. R. C. fuses, their characteristics and applications.

Unit III : Digital and Numerical Protection   6 Hrs.

Introduction, working principle , Diff. methods of Digital and Numerical protection.

Unit IV : Bus Bar   6 Hrs.

Feeder and Transmission line protection. Bus bar protection, Frame leakage protection circulating current protection and Transmission line protection using over current relays. Principles of distance relaying, choice between impedance, reactance and mho types, pilot wire and carrier pilot protection.

Unit V : Protection of Alternators and Transformers   6 Hrs.

Alternators – Stator fault, stator inter turn protection. Unbalanced load, protection (Negative phase sequence [NPS] protection)

Transformer – Use of Buccholz relay, differential protection, connection of C. T. and calculation of C.T. ratio needed for differential relaying, balanced and unbalanced restricted earth fault protection, frame leakage protection.

Generator-Transformer unit protection

Unit VI : Insulation Co-ordination and Over Current Protection   7 Hrs.

Definitions (Dry flashover voltage FOV), WEF FOV, Impulse FOV, insulation, coordinating insulation and protective devices. Basic impulse insulation (BIL), Determination of line insulation. Insulation levels of substation equipment. Lightning arrester selection and location. Modern surge diverters and Necessity of power system earthing, Method of earthing the neutral, Peterson coil, earthing of transformer.

CONTENTS

PROTECTIVE RELAYS

1.1 INTRODUCTION OF POWER SYSTEM PROTECTION

- The purpose of an Electric Power System is to generate and supply electrical energy to consumers. The power system should be designed and managed to deliver this energy to the utilization points with both reliability and economically

- The capital investment involved in power system for the generation, transmission and distribution is so great that the proper precautions must be taken to ensure that the equipment not only operates as nearly as possible to peak efficiency, but also must be protected from accidents.

- The normal path of the electric current is from the power source through copper (or aluminium) conductors in generators, transformers and transmission lines to the load and it is confined to this path by insulation. The insulation, however, may break down, either by the effect of temperature and age or by a physical accident, so that the current then follows an abnormal path generally known as Short Circuit or Fault.

- Any abnormal operating state of a power system is known as FAULT. Faults in general consist of short circuits as well as open circuits. Open circuit faults are less frequent than short circuit faults, and often they are transformed in to short circuits by subsequent events.

1.1.1 Different Types of Switchgear

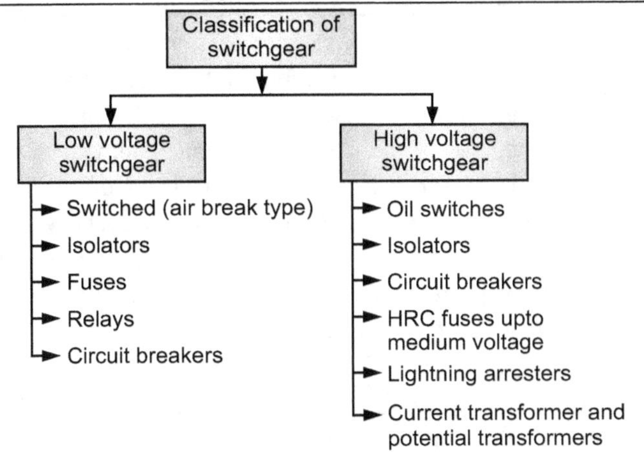

Fig. 1.1

1.1.2 Consequences of Occurrence of Faults

Faults are of Two Types

1. Short circuit fault- current
2. Open circuit fault- voltage

In terms of seriousness of consequences of a fault, short circuits are of far greater concern than open circuits, although some open circuits present some potential hazards to personnel

Classification of Short Circuited Faults

- Three phase faults (with or without earth connection)
- Two phase faults (with or without earth connection)
- Single phase to earth faults

Classification of Open Circuit Faults

- Single Phase open Circuit
- Two phase open circuit
- Three phase open circuit

Consequences

- Damage to the equipment due to abnormally large and unbalanced currents and low voltages produced by the short circuits.

- Explosions may occur in the equipments which have insulating oil, particularly during short circuits. This may result in fire and hazardous conditions to personnel and equipments.

- Individual generators with reduced voltage in a power station or a group of generators operating at low voltage may lead to loss of synchronism, subsequently resulting in islanding.

- Risk of synchronous motors in large industrial premises falling out of step and tripping out.

The general layout of a protection system may be viewed as given in the following Fig. 1.2.

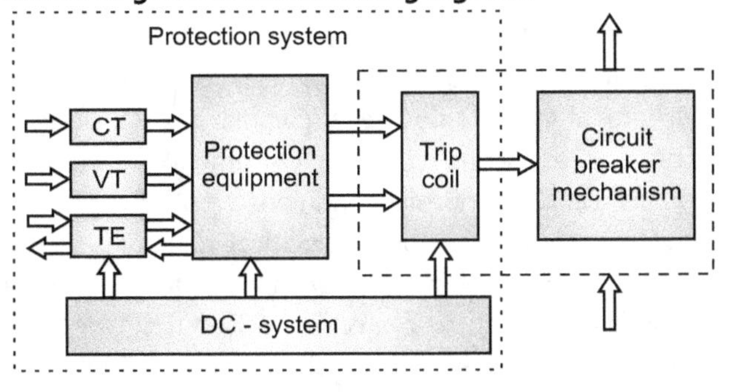

Fig. 1.2

1.2 NEED OF PROTECTIVE RELAYING IN POWER SYSTEM

- An electrical power system consists of generators, transformers, transmission and distribution lines, etc. Short circuits and other abnormal conditions often occur on a power system. The heavy current associated with short circuits is likely to cause damage to equipment if suitable protective relays and circuit breakers are not provided for the protection of each section of the power system. Short circuits are usually called faults by power engineers. Strictly speaking, the term 'fault' simply means a 'defect'. Some defects, other than short circuits, are also termed as faults. For example, the failure of conducting path due to a break in a conductor is a type of fault.

- If a fault occurs in an element of a power system, an automatic protective device is needed to isolate the faulty element as quickly as possible to keep the healthy section of the system in normal operation. The fault must be cleared within a fraction of a second. If a short circuit persists on a system for a longer, it may cause damage to some important sections of the system.

- A heavy short circuit current may cause a fire. It may spread in the system and damage a part of it. The system voltage may reduce to a low level and individual generators in a power station or groups of generators in different power stations may lose synchronism. Thus, an uncleared heavy short circuit may cause the total failure of the system.

- A protective system includes circuit breakers, transducers (CTs and PTs), and protective relays to isolate the faulty section of the power system from the healthy sections. A circuit breaker can disconnect the faulty element of the system when it is called upon to do so by the protective relay. Transducers (CTs and VTs) are used to reduce currents and voltages to lower values and to isolate protective relays from the high voltages of the power system.

- The function of a protective relay is to detect and locate a fault and issue a command to the circuit breaker to disconnect the faulty element. It is a device which senses abnormal conditions on a power system by constantly monitoring electrical quantities of the systems, which differ under normal and abnormal conditions. The basic electrical quantities which are likely to change during abnormal conditions are current, voltage, phase-angle (direction) and frequency. Protective relays utilise one or more of these quantities to detect abnormal conditions on a power system.

- Protection is needed not only against short circuits but also against any other abnormal conditions which may arise on a power system. A few examples of other abnormal conditions are over speed of generators and motors, overvoltage, under frequency, loss of excitation, overheating of stator and rotor of an alternator etc. Protective relays are also provided to detect such abnormal conditions and issue alarm signals to alert operators or trip circuit breaker.

- A protective relay does not anticipate or prevent the occurrence of a fault, rather it takes action only after a fault has occurred. However, one exception to this is the Buchholz relay, a gas actuated relay, which is used for the protection of power transformers. Sometimes, a slow breakdown of insulation due to a minor arc may take place in a transformer, resulting in the generation of heat and decomposition of the transformer's oil and solid insulation. Such a condition produces a gas which is collected in a gas chamber of the Buchholz relay. When a specified amount of gas is accumulated, the Buchholz relay operates an alarm.

- This gives an early warning of incipient faults. The transformer is taken out of service for repair before the incipient fault grows into a serious one. Thus, the occurrence of a major fault is prevented. If the gas evolves rapidly, the Buchholz relay trips the circuit breaker instantly.

- The cost of the protective equipment generally works out to be about 5% of the total cost of the system.

1.3 GENERAL IDEA ABOUT PROTECTIVE ZONE

- A power system contains generators, transformers, bus bars, transmission and distribution lines, etc. There is a separate protective scheme for each piece of equipment or element of the power system, such as generator protection, transformer protection, transmission line protection, bus bar protection, etc. Thus, a power system is divided into a number of zones for protection. A protective zone covers one or at the most two elements of a power system.

- The protective zones are planned in such a way that the entire power system is collectively covered by them, and thus, no part of the system is left unprotected. The various protective zones of a typical power system are shown in Fig. 1.3. Adjacent protective zones must overlap each other, failing which a fault on the boundary of the zones may not lie in any of the zones (this may be due to errors in the measurement of actuating quantities, etc.), and hence no circuit breaker would trip.

- Thus, the overlapping between the adjacent zones is unavoidable. If a fault occurs in the overlapping zone in a properly protected scheme, more circuit breakers than the minimum necessary to isolate the faulty element of the system would trip. A relatively low extent of overlap reduces the probability of faults in this region and consequently, tripping of too many breakers does not occur frequently.

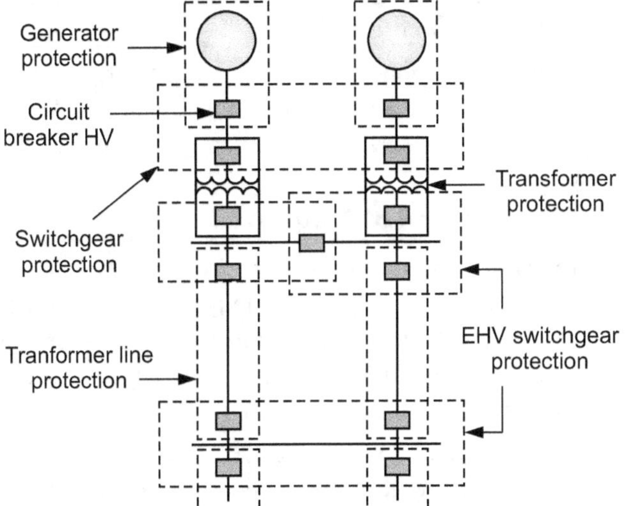

Fig. 1.3: Zones of protection

1.3.1 Primary and Back-Up Protection

It has already been explained that a power system is divided into various zones for its protection. There is a suitable protective scheme for each zone. If a fault occurs in a particular zone, it is the duty of the primary relays of that zone to isolate the faulty element. The primary relay is the first line of defense. If due to any reason, the primary relay fails to operate, there is a back-up protective scheme to clear the fault as a second line of defence.

- The causes of failures of protective scheme may be due to the failure of various elements.

- The reliability of protective scheme should at least be 95%. With proper design, installation and maintenance of the relays, circuit breakers, trip mechanisms, ac and dc wiring, etc. a very high degree of reliability can be achieved.

- The back-up relays are made independent of those factors which might cause primary relays to fail. A back-up relay operates after a time delay to give the primary relay sufficient time to operate. When a back-up relay operates, a larger part of the power system is disconnected from the power source, but this is unavoidable.

- As far as possible, a back-up relay should be placed at a different station. Sometimes, a local back-up is also used. It should be located in such a way that it does not employ components (VT, CT, measuring unit, etc.) common with the primary relays which are to be backed up. There are three types of back-up relays:
  1. Remote back-up
  2. Relay back-up
  3. Breaker back-up

1. Remote Back-up

- When back-up relays are located at a neighbouring station, they back-up the entire primary protective scheme which includes the relay, circuit breaker, VT, CT and other elements, in case of a failure of the primary protective scheme. It is the cheapest and the simplest form of back-up protection and is a widely used back-up protection for transmission lines. It is most desirable because of the fact that it will not fail due to the factors causing the failure of the primary protection.

2. Relay Back-up

- This is a kind of a local back-up in which an additional relay is provided for back-up protection. It trips the same circuit breaker if the primary relay fails and this operation takes place without delay. Though such a back-up is costly, it can be recommended where a remote back-up is not possible. For back-up relays, principles of operation that are different from those of the primary protection as desirable. They should be supplied from separate current and potential transformers.

3. Breaker Back-up

- This is also a kind of a local back-up. This type of back-up is necessary for a bus bar system where a number of circuit breakers are connected to it. When a protective relay operates in response to a fault but the circuit breaker fails to trip, the fault is treated as a bus bar fault. In such a situation, it becomes necessary that all other circuit breakers on that bus bar should trip. After a time-delay, the main relay closes the contact of a back-up relay which trips all other circuit breakers on the bus if the proper breaker does not trip within a specified time after its trip coil is energized.

1.3.2 Essential Qualities of Protection

The principal function of protective relaying is to cause the prompt removal from service of any element of the power system when it starts to operate in an abnormal manner or interfere with the effective operation of the rest of the system. In order that protective relay system may perform this function satisfactorily, it should have the following qualities :

1. Selectivity
2. Speed
3. Sensitivity
4. Reliability
5. Simplicity
6. Economy

1. **Selectivity :** It is the ability of the protective system to select correctly that part of the system in trouble and disconnect the faulty part without disturbing the rest of the system.

2. **Speed :** The relay system should disconnect the faulty section as fast as possible for the following reasons

    (i) Electrical apparatus may be damaged if they are made to carry the fault currents for a long time.

    (ii) A failure on the system leads to a great reduction in the system voltage. If the faulty section is not disconnected quickly, then the low voltage created by the fault may shut down consumers' motors and the generators on the system may become unstable.

    (iii) The high speed relay system decreases the possibility of development of one type of fault into the other more severe type.

3. **Sensitivity :** It is the ability of the relay system to operate with low value of actuating quantity. Sensitivity of a relay is a function of the volt-amperes input to the coil of the relay necessary to cause its operation. The smaller the volt-ampere input required to cause relay operation, the more sensitive is the relay. Thus, a 1 VA relay is more sensitive than a 3 VA relay. It is desirable that relay system should be sensitive so that it operates with low values of volt-ampere input.

4. **Reliability :** It is the ability of the relay system to operate under the pre-determined conditions. Without reliability, the protection would be rendered largely ineffective and could even become a liability.

5. **Simplicity :** The relaying system should be simple so that it can be easily maintained. Reliability is closely related to simplicity. The simpler the protection scheme, the greater will be its reliability.

6. **Economy :** The most important factor in the choice of a particular protection scheme is the economic aspect. Sometimes it is economically unjustified to use an ideal scheme of protection and a compromise method has to be adopted. As a rule, the protective gear should not cost more than 5% of total cost. However, when the apparatus to be protected is of utmost importance (*e.g.* generator, main transmission line etc.), economic considerations are often subordinated to reliability.

1.4 PROTECTIVE RELAYS

- A **protective relay** is a device that detects the fault and initiates the operation of the circuit breaker to ioslate the defective element from the rest of the system.

1.4.1 Classification of Protective Relays

- Protective relays can be classified in various ways depending on the technology used for their construction, their speed of operation, their generation of development, function, etc., and will be discussed in more details in the following chapters

- Classification of Protective Relays Based on Technology Protective relays can be broadly classified into the following three categories, depending on the technology they use for their construction and operation.

    1. Electromechanical relays
    2. Static relays
    3. Numerical relays or microprocessor relays

1.4.2 Principal Types of Electromagnetic Relays

There are two types of electromagnetic relays

1. Attracted armature type
2. Induction type

1. **Attracted Armature Type**

- Electromagnetic attraction relays operate by virtue of an armature being attracted to the poles of an electromagnet or a plunger being drawn into a solenoid. Such relays may be actuated by d.c. or a.c. quantities. The important types of electromagnetic attraction relays are :

    (i) Solenoid type relay

    (ii) Balanced Beam Type Relay

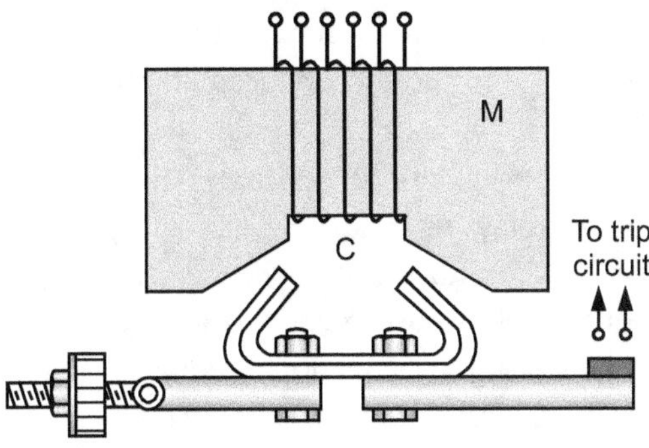

Fig. 1.4

- Fig. 1.4 above shows the schematic arrangement of an attracted armature type relay. It consists of a laminated electromagnet M carrying a coil C and a pivoted laminated armature. The armature is balanced by a counterweight and carries a pair of spring contact fingers at its free end. Under normal operating conditions, the current through the relay coil C is such that counterweight holds the armature in the position shown.

- However, when a short-circuit occurs, the current through the relay coil increases sufficiently and the relay armature is attracted upwards. The contacts on the relay armature bridge a pair of stationary contacts attached to the relay frame. This completes the trip circuit which results in the opening of the circuit breaker and, therefore, in the disconnection of the faulty circuit.

- The minimum current at which the relay armature is attracted to close the trip circuit is called pickup current. It is a usual practice to provide a number of tappings on the relay coil so that the number of turns in use and hence the setting value at which the relay operates can be varied.

(i) Solenoid Type Relay : Fig. 1.5 shows the schematic arrangement of a solenoid type relay. It consists of a solenoid and movable iron plunger arranged as shown. Under normal operating conditions, the current through the relay coil C is such that it holds the plunger by gravity or spring in the position shown. However, on the occurrence of a fault, the current through the relay coil becomes more than the pickup value, causing the plunger to be attracted to the solenoid. The upward movement of the plunger closes the trip circuit, thus opening the circuit breaker and disconnecting the faulty circuit.

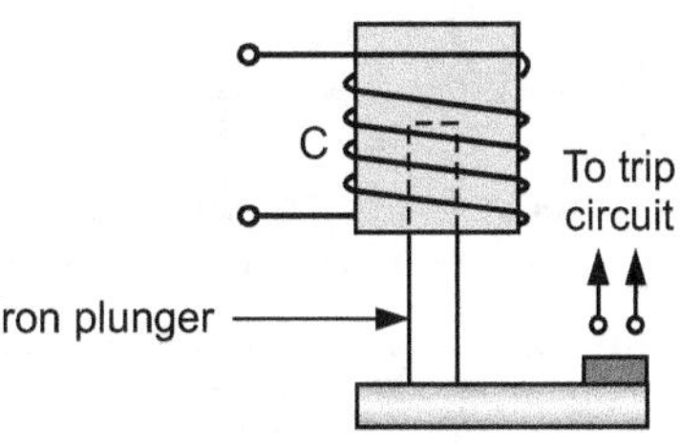

Fig. 1.5

(ii) Balanced Beam Type Relay : Fig. 1.6 shows the schematic arrangement of a balanced beam type relay. It consists of an iron armature fastened to a balance beam. Under normal operating conditions, the current through the relay coil is such that the beam is held in the horizontal position by the spring. However, when a fault occurs, the current through the relay coil becomes greater than the pickup value and the beam is attracted to close the trip circuit. This causes the opening of the circuit breaker to isolate the faulty circuit.

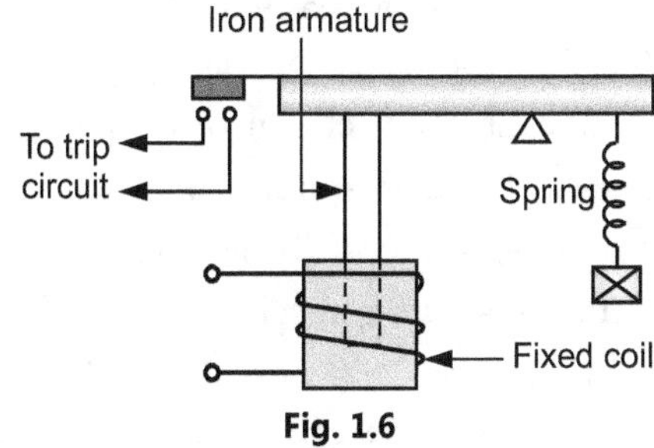

Fig. 1.6

1.4.3 Operating Principle

$$T = K_1 I_1^2 - K_2 I_2^2$$

where, $\quad$ T = Net torque

K_1, K_2 = Constants

I_1 = Current in operating coil

I_2 = Current in restraining coil

When the relay is on the verge of operation. Net torque = 0.

$$\therefore \quad K_1 I_1^2 = K_2 I_2^2$$

$$\therefore \quad \frac{I_1}{I_2} = \sqrt{\frac{K_2}{K_1}} = \text{Constant}$$

Operating Characteristics :

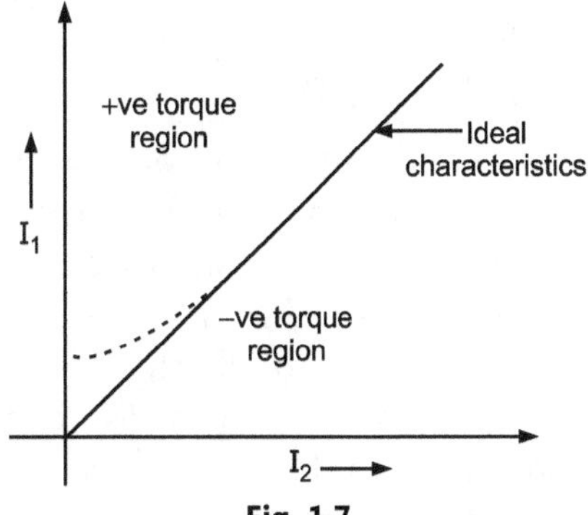

Fig. 1.7

As
$$\frac{I_1}{I_2} = \sqrt{\frac{K_2}{K_1}} = \text{constant}$$

Operating characteristic is a straight line passing through origin as shown in Fig. 1.7. When the operating current is less, the linearity of curve is not maintained, because of springs.

Advantages of Electromagnetic Relays

The various advantages of electromagnetic relays are,

- Can be used for both a.c. and d.c.

- They have last operation and fast reset

- These arc almost instantaneous. Though instantaneous, the operating time varies with current. With extra arrangements like dashpot, copper ring etc slow operating and resetting times can be obtained.

- High operating speed with operating time in few milliseconds also can be achieved.

- The pickup can he as high as 90-95% for d.c. operation and f>0 to 90% for the d.c. operation.

- Modern relays an compact, simple, reliable and robust

Disadvantages of Electromagnetic Relays

The few disadvantages of these relays are,

- The directional feature is absent.

- Due to last operation the working can be affected by the transients As transients contain d.c. as well as pulsating component, under steady stall value less than set value, the relay can operate during transients.

Applications of Electromagnetic Relays

The various applications of these relays are,

- The protection of various a.c. and d c. equipments.

- The over/under current and over/under voltage protection of various a.c. and d.c. equipments

- In the definite time lag over current and earth fault protection along with definite time lag over current relay.

- For the differential protection.

- Used as auxiliary relays in the contact systems of protective relaying schemes.

1.5 INDUCTION RELAYS

- Electromagnetic induction relays operate on the principle of induction motor and are widely used for protective relaying purposes involving a.c. quantities. They are not used with d.c. quantities owing to the principle of operation. An induction relay essentially consists of a pivoted aluminum disc placed in two alternating magnetic fields of the same frequency but displaced in time and space. The torque is produced in the disc by the interaction of one of the magnetic fields with the currents induced in the disc by the other.

- To understand the production of torque in an induction relay, refer to the elementary arrangement shown in Fig. 1.8 (i). The two a.c. fluxes ϕ_2 and ϕ_1 differing in phase by an angle α induce e.m.f.s' in the disc and cause the circulation of eddy currents i_2 and i_1 respectively. These currents lag behind their respective fluxes by 90°.

- Referring to Fig. 1.8 (ii) where the two a.c. fluxes and induced currents are shown separately for clarity, let

$$\phi_1 = \phi_{1\,max} \sin \omega t$$
$$\phi_2 = \phi_{1\,max} \sin \omega t$$

where ϕ_1 and ϕ_2 are the instantaneous values of fluxes and ϕ_2 leads ϕ_1 by an angle α. Assuming that the paths in which the rotor currents flow have negligible self-inductance, the rotor currents will be in phase with their voltages.

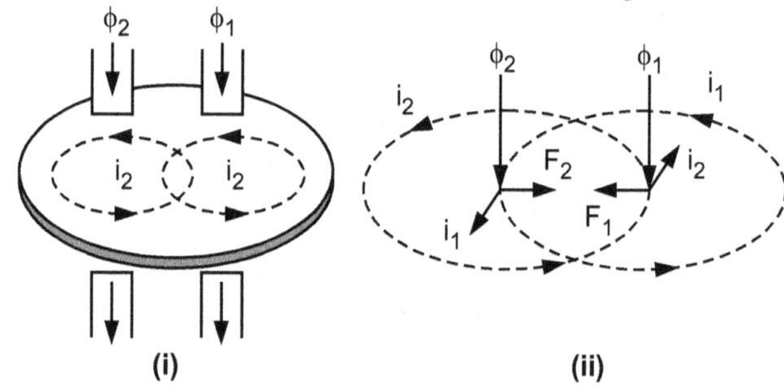

Fig. 1.8

Therefore

$$i_1 \propto \frac{d\phi_1}{dt} \propto \frac{d}{dt} (\phi_{1\,max} \sin \omega t)$$

$$\propto \phi_{1\,max} \cos \omega t$$

and

$$i_2 \propto \frac{d\phi_2}{dt} \propto \phi_{2\,max} \cos (\omega t + \alpha)$$

Now, $F_1 \propto \phi_1 i_2$ and $F_2 \propto \phi_2 i_2$

Fig. 1.8 (ii) shows that the two forces are in opposition.

∴   Net force Fat the instant considered is

$$F \propto F_2 - F_1$$
$$\propto \phi_2 i_2 - \phi_1 i_2$$
$$\propto \phi_{2max} \sin (\omega t + \alpha)\, \phi_{1max} \cos \omega t$$
$$- \phi_{1max} \sin \phi_{2max} \cos (\omega t + \alpha)$$
$$\propto \phi_{1max} \phi_{2max} [\sin (\omega t + \alpha) \cos \omega t$$
$$- \sin \omega t \cos (\omega t + \alpha)]$$
$$\propto \phi_{1max} \phi_{2max} \sin \alpha$$
$$\propto \phi_1 \phi_2 \sin \alpha \qquad \qquad \dots(1.1)$$

where ϕ_1 and ϕ_1 are the r.m.s. values of the fluxes.

The following points may be noted from exp. (1.1):

1. The greater the phase angle α between the fluxes, the greater is the net force applied to the disc. Obviously, the maximum force will be produced when the two fluxes are 90° out of phase.

2. The net force is the same at every instant. This fact does not depend upon the assumptions made in arriving at exp. (1.1).

3. The direction of net force and hence the direction of motion of the disc depends upon which flux is leading. The following three types of structures are commonly used for obtaining the phase difference in the fluxes and hence the operating torque in induction relays :

   (i)   shaded-pole structure

   (ii) watt-hour-meter or double winding structure

   (iii) induction cup structure

(i) Shaded-Pole Structure

- The general arrangement of shaded-pole structure is shown in Fig. 1.9. It consists of a pivoted aluminium disc free to rotate in the air-gap of an electromagnet. Onehalf of each pole of the magnet is surrounded by a copper band known as *shading ring.*

- The alternating flux ϕs in the shaded protion of the poles will, owing to the reaction of the current induced in the ring, lag behind the flux ϕu in the unshaded portion by an angle α. These two a.c. fluxes differing in phase will produce the necessary torque to rotate the disc. As proved earlier, the driving torque T is given by;

   given by ; $\quad T \propto \phi_s \, \phi_u \sin \alpha$

Assuming the fluxes ϕ_s and ϕ_u to be proportional to the current I in the relay coil,

$$T \propto I^2 \sin \alpha$$

This shows that driving torque is proportional to the square of current in the relay coil.

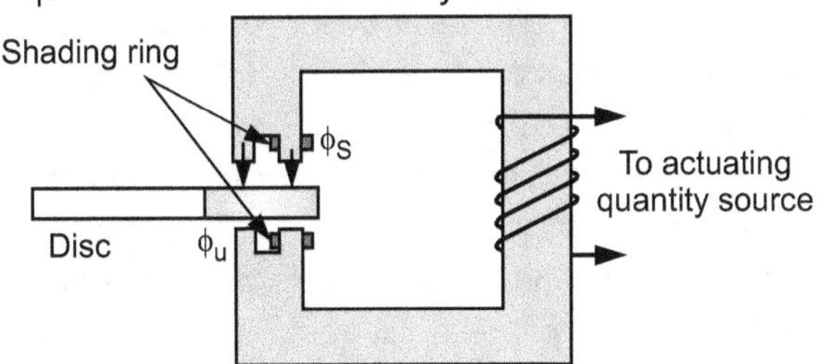

Fig. 1.9

(ii) Watthour-Meter Structure

- This structure gets its name from the fact that it is used in watthour meters. The general arrangement of this type of relay is shown in Fig. 1.10. It consists of a pivoted aluminium disc arranged to rotate freely between the poles of two electromagnets. The upper electromagnet carries two windings ; the pirmary and the secondary.

- The primary winding carries the relay current I_1 while the secondary winding is connected to the winding of the lower magnet. The primary current induces e.m.f. in the secondary and so circulates a current I_2 in it. The flux ϕ_2 induced in the lower magnet by the current in the secondary winding of the upper magnet will lag behind ϕ_1 by an angle α.

- The two fluxes ϕ_1 and ϕ_2 differing in phase by α will produce a driving torque on the disc proportional to $\phi_1 \phi_2 \sin \alpha$. An important feature of this type of relay is that its operation can be controlled by opening or closing the secondary winding circuit. If this circuit is opened, no flux can be set by the lower magnet however great the vaule of current in the pirmary winding may be and consequently no torque will be produced. Therefore, the relay can be made inoperative by opening its secondary winding circuit.

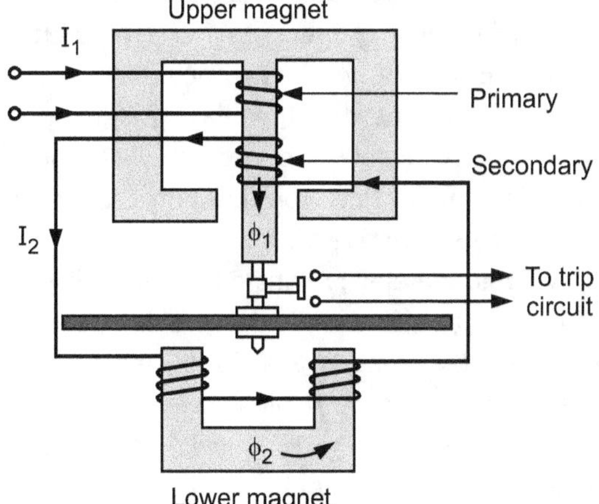

Fig. 1.10

(iii) Induction Cup Structure

- Fig. 1.11 shows the general arrangement of an induction cup structure. It most closely resembles an induction motor, except that the rotor iron is stationary, only the rotor conductor portion being free to rotate. The moving element is a hollow cylindrical rotor which turns on its axis. The rotating field is produced by two pairs of coils wound on four poles as shown.

- The rotating field induces currents in the cup to provide the necessary driving torque. If ϕ_1 and ϕ_2 represent the fluxes produced by the respective pairs of poles, then torque produced is proportional to $\phi_1 \phi_2 \sin \alpha$ where α is the phase difference between the two fluxes. A control spring and the back stop for closing of the contacts carried on an arm are attached to the spindle of the cup to prevent the continuous rotation. Induction cup structures are more efficient torque producers than either the shaded-pole or the watthour meter structures. Therefore, this type of relay has very high speed and may have an operating time less then 0.1 second.

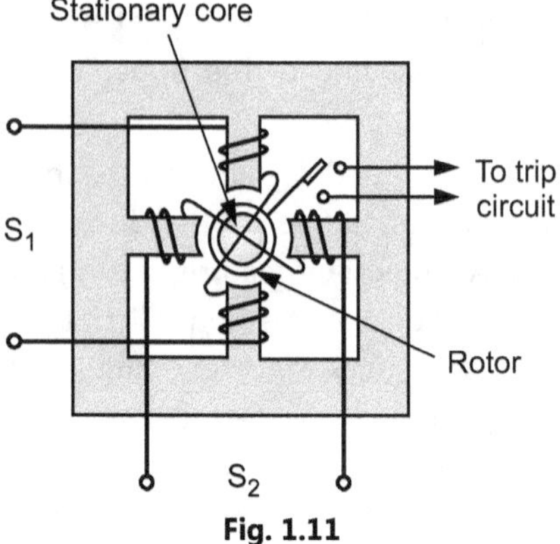

Fig. 1.11

1.5.1 Relay Timing

An important characteristic of a relay is its time of operation. By 'the time of operation' is meant length of the time from the instant when the actuating element is energised to the instant when the relay contacts are closed. Sometimes it is desirable and necessary to control the operating time of a relay. For this purpose, mechanical accessories are used with relays.

1. **Instantaneous Relay :** An instantaneous relay is one in which no intentional time delay is provided. In this case, the relay contacts are closed immediately after current in the relay coil exceeds the minimum calibrated value. Fig. 1.12 shows an instantaneous solenoid type of relay. Although there will be a short time interval between the instant of pickup and the closing of relay contacts, no intentional time delay has been added. The instantaneous relays have operating time less than 0·1 second. The instantaneous relay is effective only where the impedance between the relay and source is small compared to the protected section impedance. The operating time of instantaneous relay is sometimes expressed in cycles based on the power-system frequency e.g. one-cycle would be 1/50 second in a 50-cycle system.

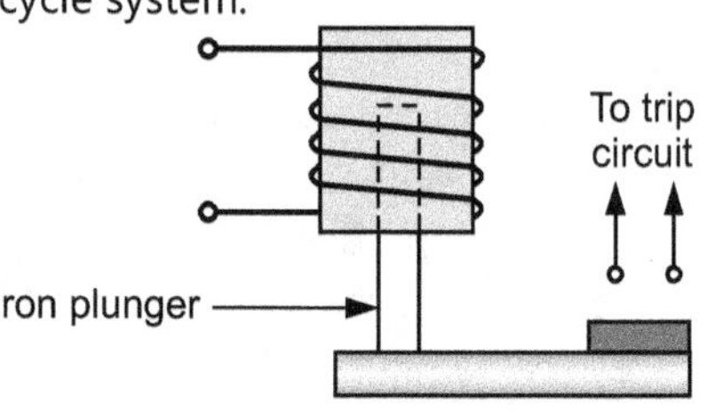

Fig. 1.12

2. **Inverse-Time Relay :** An inverse-time relay is one in which the operating time is approximately inversely proportional to the magnitude of the actuating quantity. Fig. 1.13 (a) shows the time current characteristics of an inverse current relay. At values of current less than pickup, the relay never operates. At higher values, the time of operation of the relay decreases steadily with the increase of current. The inverse-time delay can be achieved by associating mechanical accessories with relays.

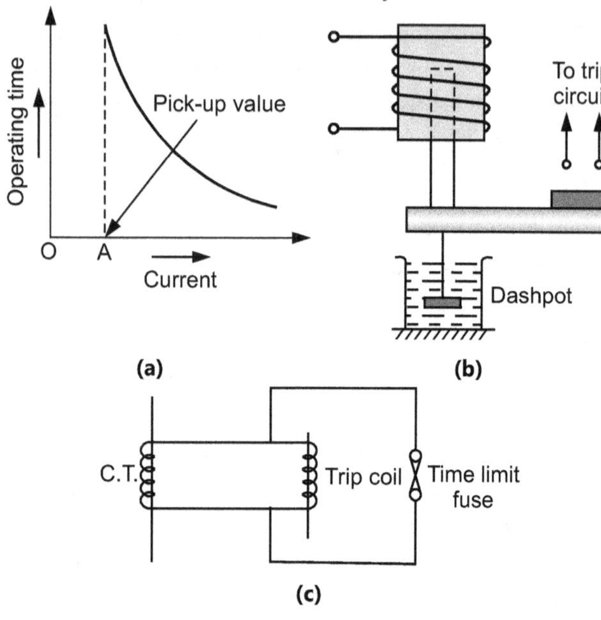

Fig. 1.13

- In an induction relay, the inverse-time delay can be achieved by positioning a permanent magnet (known as a drag magnet) in such a way that relay disc cuts the flux between the poles of the magnet. When the disc moves, currents set up in it produce a drag on the disc which slows its motion.

- In other types of relays, the inverse time delay can be introduced by oil dashpot or a time limit fuse. Fig. 1.13 (b) shows an inverse time solenoid relay using oil dashpot. The piston in the oil dashpot attached to the moving plunger slows its upward motion. At a current value just equal to the pickup, the plunger moves slowly and time delay is at a maximum. At higher values of relay current, the delay time is shortened due to greater pull on the plunger.

- The inverse-time characteristic can also be obtained by connecting a time-limit fuse in parallel with the trip coil terminals as shown in Fig. 1.13 (c).

- The shunt path formed by time-limit fuse is of negligible impedance as compared with the relatively high impedance of the trip coil. Therefore, so long as the fuse remains intact, it will divert practically the whole secondary current of CT from the trip oil. When the secondary current exceeds the current carrying capacity of the fuse, the fuse will blow and the whole current will pass through the trip coil, thus opening the circuit breaker.

- The time lag between the incidence of excess current and the tripping of the breaker is governed by the characteristics of the fuse. Careful selection of the fuse can give the desired inverse-time characteristics, although necessity for replacement after operation is a disadvantage.

3. **Definite Time Lag Relay :** In this type of relay, there is a definite time elapse between the instant of pickup and the closing of relay contacts. This particular time setting is independent of the amount of current through the relay coil ; being the same for all values of current in excess of the pickup value. It may be worthwhile to mention here that practically all inverse-time relays are also provided with definite minimum time feature in order that the relay may never become instantaneous in its action for very long overload

1.6 TERMINOLOGIES USED IN PROTECTIVE RELAYING

The various terminologies used in the protective relaying are,

Protective Relay :

- It is an electrical relay, which closes its contacts when an actuating quantity reaches a certain preset value Due to closing of contacts, relay initiates a trip circuit of circuit breaker or an alarm circuit.

Relay Time :

- It is the time between the instant of fault occurrence and the instant of closure of relay contacts

Breaker Time :

- It is the time between the instant at circuit breaker operates and opens the contacts, to the instant of extinguishing the arc completely.

Fault Clearing Time :

- The total time required between the instant of fault and the instant ot final arc interruption in the circuit breaker is fault clearing time. It is sum of the relay time and circuit breaker time.

Pickup :

- A relay is said to be picked up when it moves from the OFF' position to ON   position Thus when relay operates it is said that relay has picked up.

Pickup Value :

- It is the minimum value of an actuating quantity at which relay starts operating. In most of the relays actuating quantity is current m the relay coil and pickup value of current is indicated along with the relay.

Dropout or Reset :

- A relay is said to dropout or reset when it comes back to original position ie when relay contacts open from its closed position. The value of an actuating quantity current or voltage below which the relay resets is called reset value of that relay

Time Delay :

- The time taken by relay to operate after it has sensed tlu- fault is called time delay of relay. Some relays are instantaneous while in some relays intentionally a time delay is provided.

Sealing Relays or Holding Relays :

- The relay contacts are designed for light weight and hence they are therefore very delicate When the protectrve relay closes its contacts, it is relieved from other duties such as time lag, tripping etc. These duties are performed by auxiliary relays which are also called sealing relays or holding relays.

Current Setting :

- It is often desirable to adjust the pick-up current to any required value. This is known as current setting and is usually achieved by the use of tappings on the relay operating coil. The taps are brought out to a plug bridge as shown in Fig. 1.14. The plug bridge permits to alter the number of turns on the relay coil.

- This changes the torque on the disc and hence the time of operation of the relay. The values assigned to each tap are expressed in terms of percentage full-load rating of C.T. with which the relay is associated and represents the value above which the disc commences to rotate and finally closes the trip circuit.

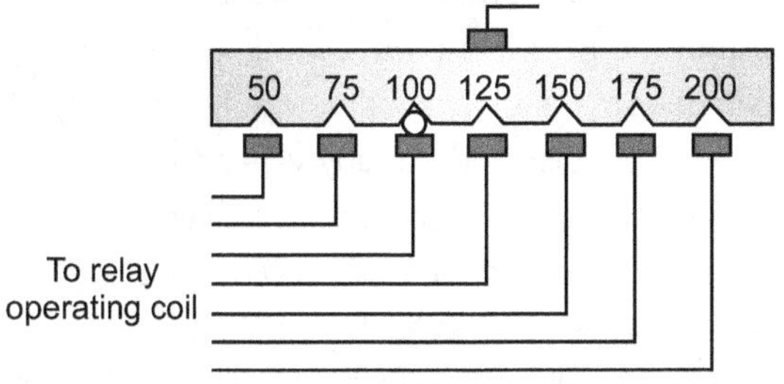

Fig. 1.14

$\therefore$ Pick-up current = Rated secondary current of C.T.
$\times$ Current setting

- For example, suppose that an overcurrent relay having current setting of 125% is connected to a supply circuit through a current transformer of 400/5. The rated secondary current of C.T. is 5 amperes. Therefore, the pick-up value will be 25% more than 5 A *i.e.* 5 × 1·25 = 6·25 A. It means that with above current setting, the relay will actually operate for a relay coil current equal to or greater than 6·25 A.

- The current plug settings usually range from 50% to 200% in steps of 25% for overcurrent relays and 10% to 70% in steps of 10% for earth leakage relays. The desired current setting is obtained by inserting a plug between the jaws of a bridge type socket at the tap value required.

Plug-Setting Multiplier (P.S.M.) :

- It is the ratio of fault current in relay coil to the pick-up current i.e.

$$= \frac{\text{Fault current in relay coil}}{\text{Rated secondary current of CT} \times \text{Current setting}}$$

$$\text{PSM} = \frac{\text{Fault current in relay coil}}{\text{Pick-up current}}$$

Time-Setting Multiplier

- A relay is generally provided with control to adjust the time of operation. This adjustment is known as time-setting multiplier. The time-setting dial is calibrated from 0 to 1 in steps of 0.05 sec (see Fig. 1.15). These figures are multipliers to be used to convert the time derived from time/P.S.M. curve into the actual operating time. Thus if the time setting is 0·1 and the time obtained from the time/P.S.M. curve is 3 seconds, then actual relay operating time = 3 × 0·1 = 0·3 second.

- For instance, in an induction relay, the time of operation is controlled by adjusting the amount of travel of the disc from its reset position to its pickup position. This is achieved by the adjustment of the position of a movable backstop which controls the travel of the disc and thereby varies the time in which the relay will close its contacts for given values of fault current. A so-called "time dial" with an evenly divided scale provides this adjustment. The acutal time of operation is calculated by multiplying the time setting multiplier with the time obtained from time/P.S.M. curve of the relay.

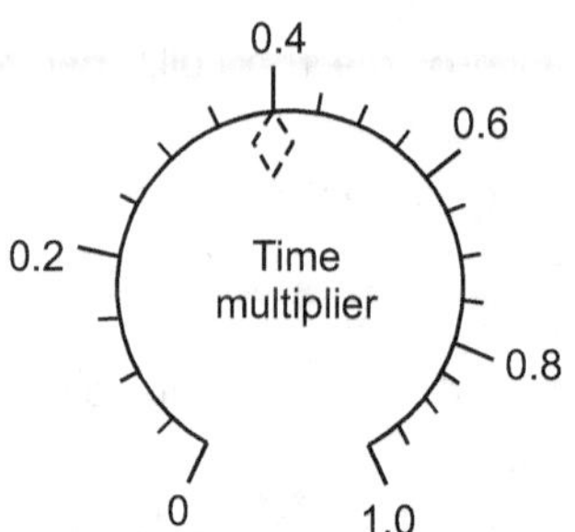

Fig. 1.15

Time/P.S.M. Curve :

- Fig. 1.16 shows the curve between time of operation and plug setting multiplier of a typical relay. The horizontal scale is marked in terms of plug-setting multiplier and represents the number of times the relay current is in excess of the current setting. The vertical scale is marked in terms of the time required for relay operation. If the P.S.M. is 10, then the time of operation (from the curve) is 3 seconds.

- The actual time of operation is obtained by multiplying this time by the time-setting multiplier. It is evident from Fig. 1.16 that for lower values of overcurrent, time of operation varies inversely with the current but as the current approaches 20 times full-load value, the operating time of relay tends to become constant. This feature is necessary in order to ensure discrimination on very heavy fault currents flowing through sound feeders.

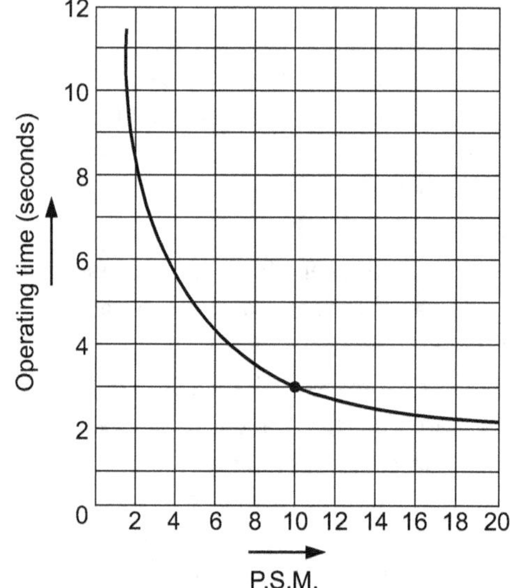

Fig. 1.16

Trip Circuit :

- The opening operation of circuit breaker is controlled by a circuit which consists of trip coil, relay contact-, auxiliary switch, battery supply etc which is called trip circuit.

Earth Fault :

- The fault involving earth i.e. called earth fault The examples of earth fault are single line to ground fault, double line to ground fault etc.

Phase Fault :

- The fault which does not involve earth is called phase fault The example is line to line fault.

Protective Scheme :

- The combination of various protective systems covering a particular protective /one for a particular equipment is called protective scheme. For example a generator may be provided with protective systems like overcurrent, differential, earth fault etc. The combination of all these systems is called generator protective scheme.

Protective System :

- The combination of circuit breakers, trip circuits, CT. and other protective relaying equipments is called protective system.

Unit Protection :

- A protective system in which the protection zone is clearly defined by the C.T. boundaries is called unit protection. Such systems work for internal faults only.

Reach :

- The limiting distance in which protective system responds to the faults is called reach of the protective system. The operation beyond the set distance is called over-reach while failure of distance relay within set distance is called under-reach.

1.7 CALCULATION OF RELAY OPERATING TIME

In order to calculate the actual relay operating time, the following things must be known :

1. Time/P.S.M. curve
2. Current setting
3. Time setting
4. Fault current
5. Current transformer ratio

The procedure for calculating the actual relay operating time is as follows :

(i) Convert the fault current into the relay coil current by using the current transformer ratio.

(ii) Express the relay current as a multiple of current setting i.e. calculate the P.S.M.

(iii) From the Time/P.S.M. curve of the relay, read off the time of operation for the calculated P.S.M.

(iv) Determine the actual time of operation by multiplying the above time of the relay by time setting multiplier in use.

1.8 FUNCTIONAL RELAY TYPES

- Most of the relays in service on power system today operate on the principle of electromagnetic attraction or electromagnetic induction. Regardless of the principle involved, relays are generally classified according to the function they are called upon to perform in the protection of elelctric power circuits.

- For example, a relay which recognizes over current in a circuit (i.e. current greater than that which can be tolerated) and initiates corrective measures would be termed as an over current relay irrespective of the relay design. Similarly an overvoltage relay is one which recognises overvoltage in a circuit and initiates the corrective measures. Although there are several types of special function relays, only the following important types will be discussed in this chapter :

1. Induction type overcurrent relays
2. Induction type reverse power relays
3. Distance relays
4. Differential relays
5. Translay scheme

1.8.1 Induction Type Over Current Relay (Non-Directional)

This type of relay works on the induction principle and initiates corrective measures when current in the circuit exceeds the predetermined value. The actuating source is a current in the circuit supplied to the relay from a current transformer. These relays are used on a.c. circuits only and can operate for fault current flow in either direction.

Constructional Details :

- Fig. 1.17 shows the important constructional details of a typical nondirectional induction type overcurrent relay. It consists of a metallic (aluminium) disc which is free to rotate in between the poles of two electromagnets. The upper electromagnet has a primary and a secondary winding. The primary is connected to the secondary of a C.T. in the line to be protected and is tapped at intervals. The tappings are connected to a plug-setting bridge by which the number of active turns on the relay operating coil can be varied, thereby giving the desired current setting.

- The secondary winding is energised by induction from primary and is connected in series with the winding on the lower magnet. The controlling torque is provided by a spiral spring. The spindle of the disc carries a moving contact which bridges two fixed contacts (connected to trip circuit) when the disc rotates through a pre-set angle. This angle can be adjusted to any value between 0° and 360°. By adjusting this angle, the travel of the moving contact can be adjusted and hence the relay can be given any desired time setting.

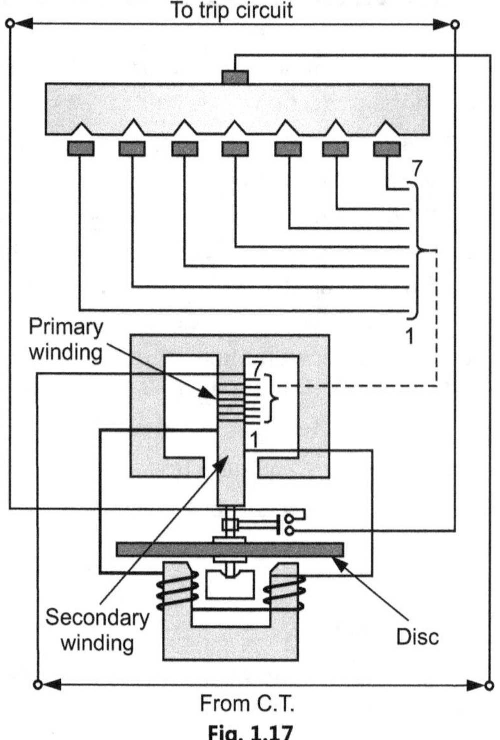

Fig. 1.17

Operation :

- The driving torque on the aluminium disc is set up due to the induction principle. This torque is opposed by the restraining torque provided by the spring.
- Under normal operating conditions, restraining torque is greater than the driving torque produced by the relay coil current. Therefore, the aluminium disc remains stationary. However, if the current in the protected circuit exceeds the pre-set value, the driving torque becomes greater than the restraining torque. Consequently, the disc rotates and the moving contact bridges the fixed contacts when the disc has rotated through a pre-set angle. The trip circuit operates the circuit breaker which isolates the faulty section.

1.8.2 Induction Type Directional Power Relay

This type of relay operates when power in the circuit flows in a specific direction. Unlike a nondirectional overcurrent relay, a directional power relay is so designed that it obtains its operating torque by the interaction of magnetic fields derived from both voltage and current source of the circuit it protects. Thus this type of relay is essentially a wattmeter and the direction of the torque set up in the relay depends upon the direction of the current relative to the voltage with which it is associated.

Constructional Details :

- Fig. 1.18 (a) shows the essential parts of a typical induction type directional power relay. It consists of an aluminum disc which is free to rotate in between the poles of two electromagnets. The upper electromagnet carries a winding (called potential coil) on the central limb which is connected through a potential transformer (P.T.) to the circuit voltage source. The lower electromagnet has a separate winding (called current coil) connected to the secondary of C.T. in the line to be protected.

- The current coil is provided with a number of tappings connected to the plugsetting bridge (not shown for clarity). This permits to have any desired current setting. The restraining torque is provided by a spiral spring. The spindle of the disc carries a moving contact which bridges two fixed contacts when the disc has rotated through a pre-set angle. By adjusting this angle, the travel of the moving disc can be adjusted and hence any desired time-setting can be given to the relay.

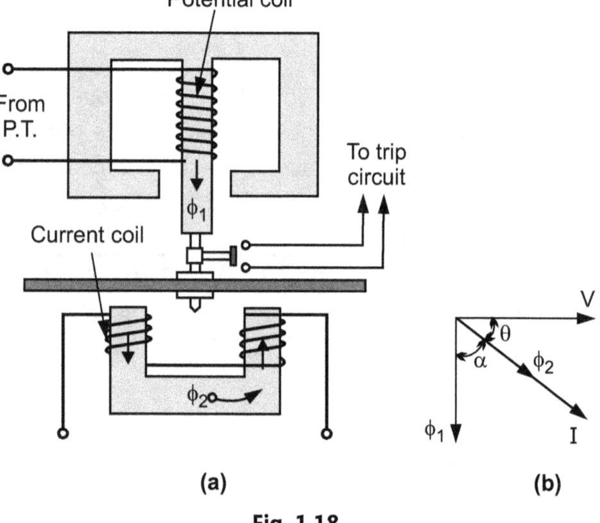

Fig. 1.18

Operation :

- The flux ϕ_1 due to current in the potential coil will be nearly 90° lagging behind the applied voltage V. The flux ϕ_2 due to current coil will be nearly in phase with the operating current I [See vector diagram in Fig. 1.18 (b)]. The interaction of fluxes ϕ_1 and ϕ_2 with the eddy currents induced in the disc produces a driving torque given by :

$$T \propto \phi_1 \phi_2 \sin \alpha$$

Since $\phi_1 \propto V$, $\phi_2 \propto I$ and $\alpha = 90 - \theta$

$$T \propto VI \sin (90 - \theta)$$

$$\propto VI \cos \theta$$

$$\propto \text{Power in the circuit}$$

It is clear that the direction of driving torque on the disc depends upon the direction of power flow in the circuit to which the relay is associated. When the power in the circuit flows in the normal direction, the driving torque and the restraining torque (due to spring) help each other to turn away the moving contact from the fixed contacts. Consequently, the relay remains inoperative. However, the reversal of current in the circuit reverses the direction of driving torque on the disc. When the reversed driving torque is large enough, the disc rotates in the reverse direction and the moving contact closes the trip circuit. This causes the operation of the circuit breaker which disconnects the faulty section.

1.8.3 Induction Type Directional Overcurrent Relay

The directional power relay discussed above is unsuitable for use as a directional protective relay under short-circuit conditions. When a short-circuit occurs, the system voltage falls to a low value and there may be insufficient torque developed in the relay to cause its operation. This difficulty is overcome in the directional overcurrent relay which is designed to be almost independent of system voltage and power factor.

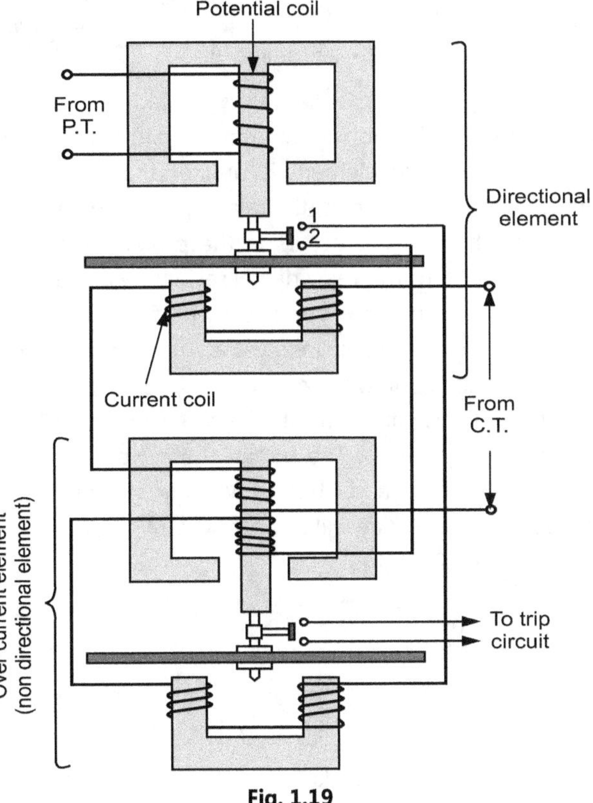

Fig. 1.19

Constructional Details:

Fig. 1.19 shows the constructional details of a typical induction type directional ovecurrent relay. It consists of two relay elements mounted on a common case *viz.*

1. Directional element and

2. Non-directional element.

1. **Directional Element:** It is essentially a directional power relay which operates when power flows in a specific direction. The potential coil of this element is connected through a potential transformer (P.T.) to the system voltage. The current coil of the element is energised through a C.T. by the circuit current. This winding is carried over the upper magnet of the non-directional element. The trip contacts (1 and 2) of the directional element are connected in series with the secondary circuit of the overcurrent element. Therefore, the latter element cannot start to operate until its secondary circuit is completed. In other words, the directional element must operate first (*i.e.* contacts 1 and 2 should close) in order to operate the overcurrent element.

2. **Non-Directional Element :** It is an overcurrent element similar in all respects to a non-directional overcurrent relay. The spindle of the disc of this element carries a moving contact which closes the fixed contacts (trip circuit contacts) after the operation of directional element. It may be noted that plug-setting bridge is also provided in the relay for current setting but has been omitted in the figure for clarity and simplicity. The tappings are provided on the upper magnet of overcurrent element and are connected to the bridge.

Operation : Under normal operating conditions, power flows in the normal direction in the circuit protected by the relay. Therefore, directional power relay (upper element) does not operate, thereby keeping the overcurrent element (lower element) unenergised. However, when a short-circuit occurs, there is a tendency for the current or power to flow in the reverse direction. Should this happen, the disc of the upper element rotates to bridge the fixed contacts 1 and 2. This completes the circuit for overcurrent element. The disc of this element rotates and the moving contact attached to it closes the trip circuit. This operates the circuit breaker which isolates the faulty section. The two relay elements are so arranged that final tripping of the current controlled by them is not made till the following conditions are satisfied :

(i) current flows in a direction such as to operate the directional element.

(ii) current in the reverse direction exceeds the pre-set value.

(iii) excessive current persists for a period corresponding to the time setting of overcurrent element.

1.9 DISTANCE OR IMPEDANCE RELAYS

- The operation of the relays discusssed so far depended upon the magnitude of current or power in the protected circuit. However, there is another group of relays in which the operation is governed by the ratio of applied voltage to current in the protected circuit. Such relays are called distance or impedance relays. In an impedance relay, the torque produced by a current element is opposed by the torque produced by a voltage element. The relay will operate when the ratio V/I is less than a predetermined value. Fig. 1.20 illustrates the basic principle of operation of an impedance relay.

- The voltage element of the relay is excited through a potential transformer (P.T.) from the line to be protected. The current element of the relay is excited from a current transformer (C.T.) in series with the line. The portion AB of the line is the protected zone. Under normal operating conditions, the impedance of the protected zone is Z_L. The relay is so designed that it closes its contacts whenever impedance of the protected section falls below the pre-determined value i.e. Z_L in this case.

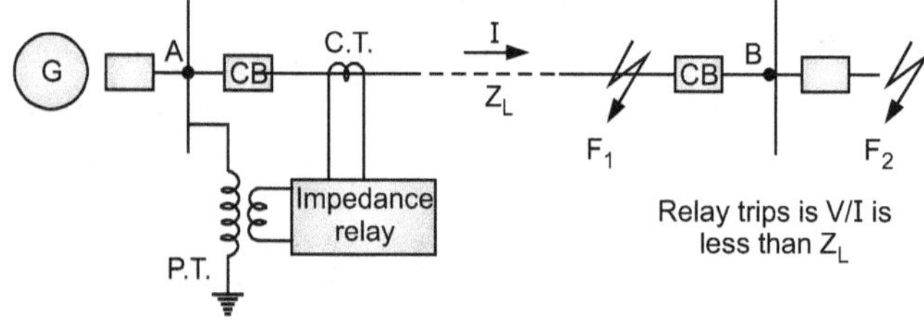

Fig. 1.20

Now suppose a fault occurs at point F_1 in the protected zone. The impedance Z (= V/I) between the point where the relay is installed and the point of fault will be less than Z_L and hence the relay operates. Should the fault occur beyond the protected zone (say point F_2), the impedance Z will be greater than Z_L and the relay does not operate.

Types : A distance or impedance relay is essentially an ohmmeter and operates whenever the impedance of the protected zone falls below a pre-determined value. There are two types of distance relays in use for the protection of power supply, namely ;

1. Definite-distance relay which operates instantaneously for fault upto a pre-determined distance from the relay.

2. Time-distance relay in which the time of operation is proportional to the distance of fault from the relay point. A fault nearer to the relay will operate it earlier than a fault farther away from the relay. It may be added here that the distance relays are produced by modifying either of two types of basic relays; the balance beam or the induction disc.

1. Definite – Distance Type Impedance Relay

- Fig. 1.21 shows the schematic arrangement of a definite-distance type impedance relay. It consists of a pivoted beam F and two electromagnets energised respectively by a current and voltage transformer in the protected circuit. The armatures of the two electromagnets are mechanically coupled to the beam on the opposite sides of the fulcrum. The beam is provided with a bridging piece for the trip contacts. The relay is so designed that the torques produced by the two electromagnets are in the opposite direction.

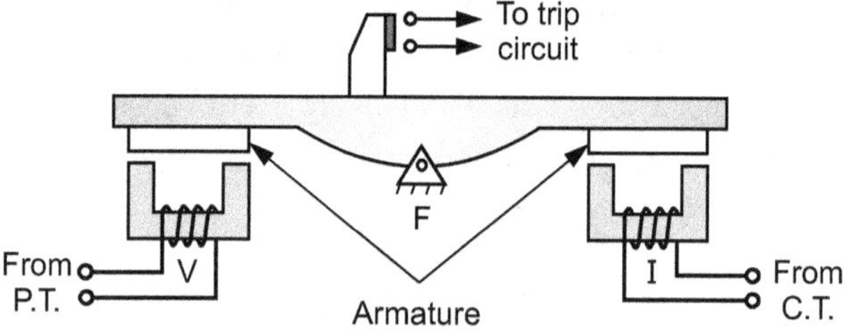

Fig. 1.21

Operation : Under normal operating conditions, the pull due to the voltage element is greater than that of the current element. Therefore, the relay contacts remain open. However, when a fault occurs in the protected zone, the applied voltage to the relay decreases whereas the current increases. The ratio of voltage to current (i.e. impedance) falls below the pre-determined value. Therefore, the pull of the current element will exceed that due to the voltage element and this causes the beam to tilt in a direction to close the trip contacts. The pull of the current element is proportional to I^2 and that of voltage element to V^2. Consequently, the relay will operate when

$$k_1 V^2 < k_2 I^2$$

or

$$\frac{V^2}{I^2} < \frac{k_2^2}{k^2}$$

or

$$\frac{V}{I} = \sqrt{\frac{k_2^2}{k_1^2}}$$

or
$$z = \sqrt{\frac{k_2^2}{k_1^2}}$$

The value of the constants k_1 and k_2 depends upon the ampere-turns of the two electromagnets.

By providing tappings on the coils, the setting value of the relay can be changed.

2. Time-Distance Impedance Relay

A time-distance impedance relay is one which automatically adjusts its operating time according to the distance of the relay from the fault point *i.e.*

Operating time,

$$T \propto \frac{V}{I}$$

$$\propto Z$$

$$\propto distance$$

Construction : Fig. 1.22 shows the schematic arrangement of a typical induction type time distance impedance relay. It consists of a current driven induction element similar to the double winding type induction over current relay .The spindle carrying the disc of this element is connected by means of a spiral spring coupling to a second spindle which carries the bridging piece of the relay trip contacts. The bridge is normally held in the open position by an armature held against the pole face of an electromagnet excited by the voltage of the circuit to be protected.

Operation : Under normal load conditions, the pull of the armature is more than that of the induction element and hence the trip circuit contacts remain open. However, on the occurence of a short-circuit, the disc of the induction current element starts to rotate at a speed depending upon the operating current. As the rotation of the disc proceeds, the spiral spring coupling is wound up till the tension of the spring is sufficient to pull the armature away from the pole face of the voltage-excited magnet. Immediately this occurs, the spindle carrying the armature and bridging piece moves rapidly in response to the tension of the spring and trip contacts are closed. This opens the circuit breaker to isolate the faulty section. The speed of rotation of the disc is approximately proportional to the operating current, neglecting the effect of control spring. Also the time of operation of the relay is directly proportional to the pull of the voltage-excited magnet and hence to the line voltage V at the point where the relay is connected. Therefore, the time of operation of relay would vary as V/I i.e. as Z or distance.

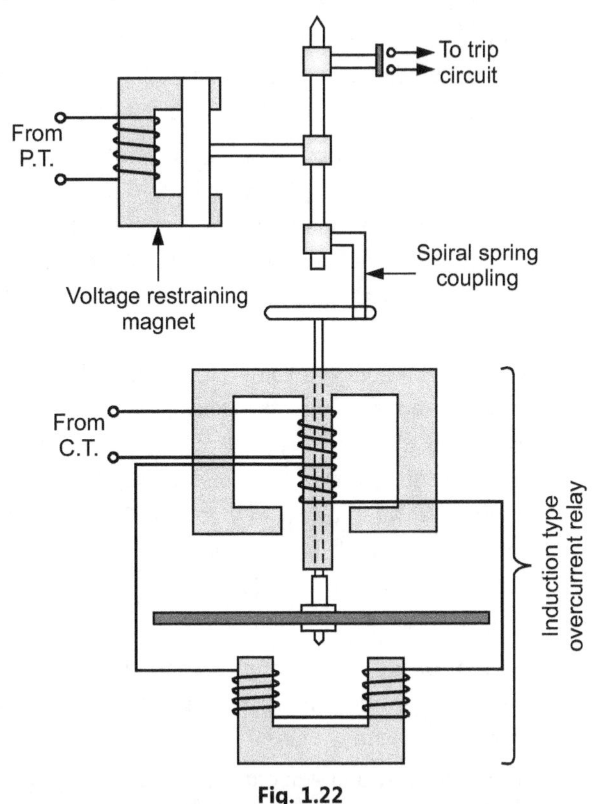

Fig. 1.22

1.10 REACTANCE RELAY

In this relay the operating torque is obtained by current while the restraining torque due to a current-voltage directional relay. The overcurrent element develops the positive torque and directional unit produces negative torque. Thus the reactance relay is an overcurrent relay with the directional restraint The directional element is so designed that the maximum torque angle is 90.

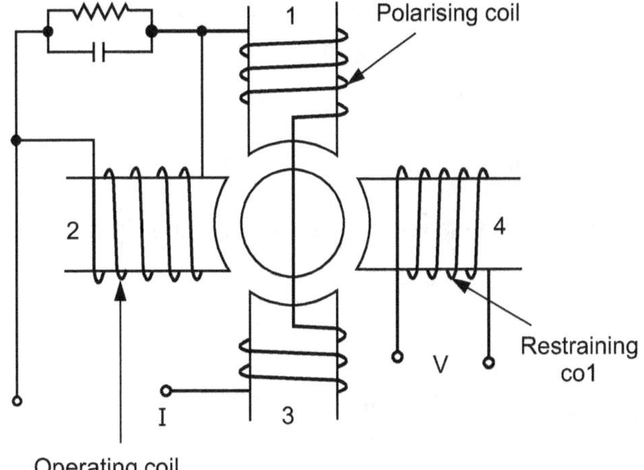

Fig. 1.23 (a) : Schematic arrangement of reactance relay

Construction :

- The structure used for the reactance relay can be of induction cup type relay   It consists   a four pole structure. It has operating coil, polarizing coil and a restraining coil The schematic arrangement of coils for the reactance relay is shown in the Fig. 1.23 (a).

- The current I flows from pole 1, through iron core stacking to lower pole 3. The winding on pole 4 is fed from voltage V. The operating torque is produced by interaction of fluxes due to the windings carrying current colls i-e interaction of fluxes produced by poles 1, 2 and 3.

- While the restraining torque is developed due to interaction of fluxes due to the poles 1, 3 and 4. Hence the operating torque is proportional to the square of the current (I^2) while the restraining torque is proportional to the product of V and I (VI). The desired maximum torque angle is obtained with the help of RC circuit, shown in the Fig. 1.23 (a).

Torque Equation

- The driving torque is proportional to the square of the current while the restraining torque is proportional to the product of V and I.

- Hence the net torque neglecting the effect of spring is given by,

$$T = K_1 I^2 - K_2 VI \cos(\theta - t)$$

At the balance point net torque is zero,

$$0 = K_1 I^2 - K_2 VI \cos(\theta - t)$$

$$\therefore \quad K_1 I^2 = K_2 VI \cos(\theta - t)$$

$$\therefore \quad K_1 = K_2 \frac{V}{I} \cos(\theta - t)$$

$$\therefore \quad K_1 = K_2 Z \cos(\theta - t)$$

Adding capacitor, the torque angle is adjusted as 90°,

$$\therefore \quad K_1 = K_2 Z \cos(\theta - 90°)$$

$$\therefore \quad K_1 = K_2 Z \sin\theta$$

$$\therefore \quad Z \sin\theta = \frac{K_1}{K_2}$$

Consider an impedance triangle shown in the Fig. 1.23 (b).

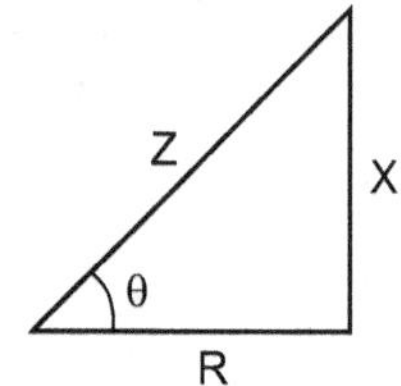

Fig. 1.23 (b)

$$Z \sin\theta = X = \text{reactance}$$

$$Z \cos\theta = R = \text{reactance}$$

$$\therefore \quad x = \frac{K_1}{K_2} = \text{constant}$$

Thus the relay operates on the reactance only. The constant X means a straight line parallel to X-axis, on R-X diagram. For the operation of the relay, the reactance seen by the relay should be smaller than the reactance for which the relay is designed.

1.10.1 Operating Characteristics

The operating characteristics to such relay is a straight line parallel to tin x-axis i.e. R-axis on R-X diagram. All the impedance vectors have their tips lying ton the straight line representing constant reactance. The resistance component of the impedance has no effect on operation of the relay. It responds only to the reactance component of the impedance. The characteristics is shown in the Fig. 1.24.

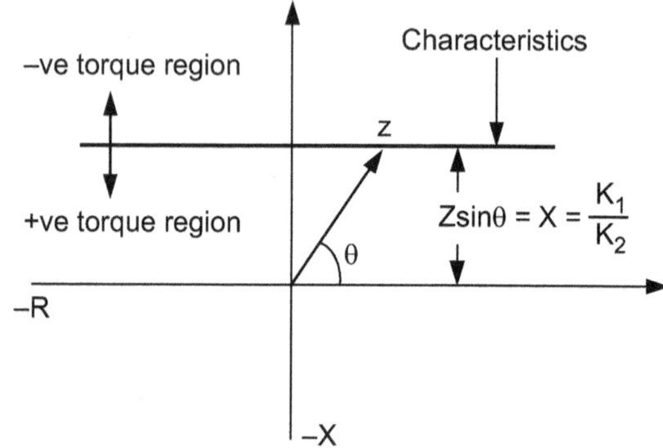

Fig. 1.24 : Operating characteristics of reactance realy

The relay will operate for all the impedances whose heads lie below the operating characteristics, whether below or above the R-axis.

Disadvantages

This relay as can be seen from the characteristics is a nondirectional relay. This will not be able discriminate when used on transmission line, whether the fault has taken place in the section where relay is located or it has taken place in the adjoining section. It is not possible to use a directional relay of the type used with bask impedance relay because in that case the relay will operate even under normal load conditions if the system is operating at or near unity p.f. conditions. The reactance relay with directional feature is called mho relay or admittance relay.

1.11 MHO RELAY OR ADMITTANCE RELAY

Mho Relay or Admittance Relay

In the impedance relay a separate unit is required to make it directional while the same unit can not be used to make a reactance relay with directional feature. The mho relay is made inherently directional by adding a voltage winding called polarizing winding. This relay works on the measurement of admittance Y_{ZO}. This relay is also called angle impedance relay.

Construction

- This relay also uses an induction cup type structure. It also has an operating coil polarizing coil and restraining coil. The schematic arrangement of all the coils is shown in the Fig. 1.25.

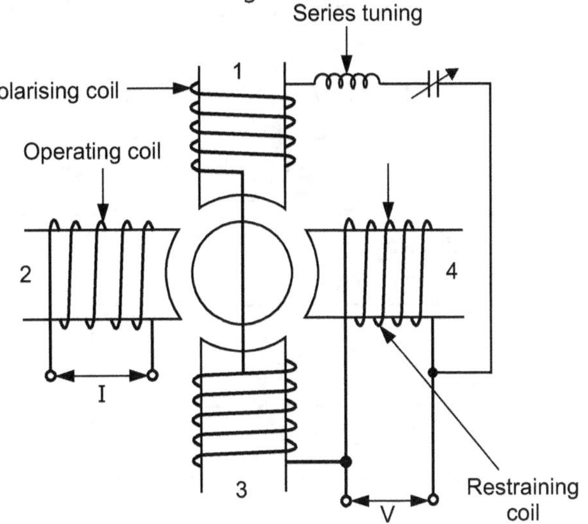

Fig. 1.25

- In this relay the operating torque is obtained by V and 1 element while the restraining torque is obtained by a voltage element. Tims an admittance relay is a voltage restrained directional relay.
- The operating torque is produced by the interaction of the fluxes due to the windings carried by the poles 1, 2 and 3. While the restraining torque is produced by the interaction of the fluxes due to the windings carried by the poles 1, 3 and 4.
- Thus the restraining torque is proportional to the square of the voltage (V) while the operating torque is proportional to the product of voltage and current (VI). The torque angle is adjusted using series tuning circuit.

Torque Equation

The operating torque is proportional to VI while restraining torque is proportional to VI while restraining torque is proportional to V^2. Hence net is given by,

$$T = K_1 V I \cos(\theta - \tau) - K_2 V^2 - K_3$$

where K_3 = control spring effect

Generally control spring effect is neglected ($K_3 = 0$).

And at balance net torque is also zero.

$$\therefore \quad 0 = K_1 VI \cos(\theta - \tau) - K_2 V^2$$

$$\therefore \quad K_1 VI \cos(\theta - \tau) = K_2 V^2$$

$$\therefore \quad K_1 \cos(\theta - \tau) = K_2 \frac{V^2}{VI}$$

$$\therefore \quad K_1 \cos(\theta - \tau) = K_2 \frac{V}{I}$$

$$\therefore \quad Z = \frac{K_1}{K_2} \cos(\theta - \tau)$$

This is the equation of a circle having diameter $\frac{K_1}{K_2}$ is the ohmic setting of the setting of this relay.

Operating Characteristics

As seen from the torque equation, the characteristics of this relay is a circle passing through origin with diameter as $\frac{K_1}{K_2}$.

Let $\frac{K_1}{K_2} = Z_R$ = ohmic setting of realy

= diameter

The circle is shown in the Fig. 1.26.

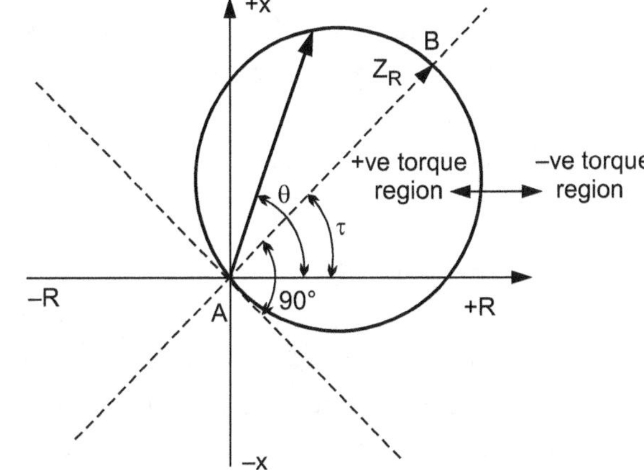

Fig. 1.26

The relay operates when the impedance seen by the relay falls within this circle. Consider two lines AB and AC with mho relay located at the point A. The relay will operate for the faults occurring in the section AB only and not for the faults occurring in the section AC. This shows that this relay is inherently directional without any additional directional unit required. The-angle τ can be adjusted to be 45° 60°, 75° and so on. This angle is maximum torque angle. The sotting of 45° is used for high voltage (33 or 11 kV) distribution lines, the setting of 60° is use for (66 or 132 kV) lines while the setting of 75° is used for (275 and 400 kV) lines.

1.12 DIFFERENTIAL RELAYS

- Most of the relays discussed so far relied on excess of current for their operation. Such relays are less sensitive because they cannot make correct distinction between heavy load conditions and minor fault conditions. In order to overcome this difficulty, differential relays are used.
- A differential relay is one that operates when the phasor difference of two or more similar electrical quantities exceeds a pre-determined value. Thus a

current differential relay is one that compares the current entering a section of the system with the current leaving the section. Under normal operating conditions, the two currents are equal but as soon as a fault occurs, this condition no longer applies.

- The difference between the incoming and outgoing currents is arranged to flow through the operating coil of the relay. If this differential current is equal to or greater than the pickup value, the relay will operate and open the circuit breaker to isolate the faulty section. It may be noted that almost any type of relay when connected in a particular way can be made to operate as a differential relay. In other words, it is not so much the relay construction as the way the relay is connected in a circuit that makes it a differential relay. There are two fundamental systems of differential or balanced protection viz.

1. Current balance protection

2. Voltage balance protection

1. Current Differential Relay

- Fig. 1.27 shows an arrangement of an overcurrent relay connected to operate as a differential relay. A pair of identical current transformers are fitted on either end of the section to be protected (alternator winding in this case). The secondaries of CT's are connected in series in such a way that they carry the induced currents in the same direction. The operating coil of the overcurrent relay is connected across the CT secondary circuit. This differential relay compares the current at the two ends of the alternator winding.

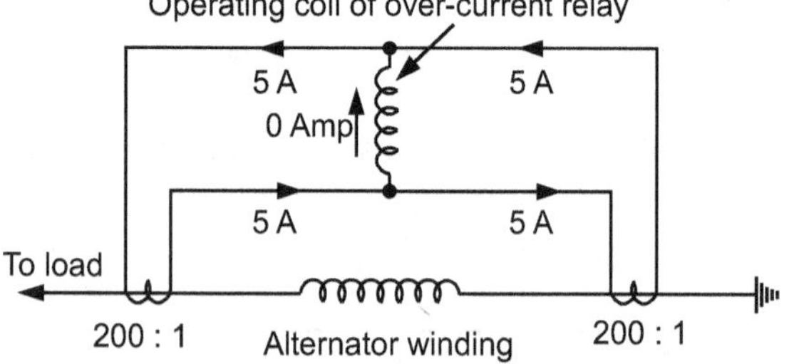

Fig. 1.27

- Under normal operating conditions, suppose the alternator winding carries a normal current of 1000 A. Then the currents in the two secondaries of CT's are equal [See Fig. 1.27]. These currents will merely circulate between the two CT's and no current will flow through the differential relay. Therefore, the relay remains inoperative. If a ground fault occurs on the

alternator winding as shown in Fig. 1.28 (a), the two secondary currents will not be equal and the current flows through the operating coil of the relay, causing the relay to operate. The amount of current flow through the relay will depend upon the way the fault is being fed.

(i) If some current (500 A in this case) flows out of one side while a larger current (2000 A) enters the other side as shown in Fig. 1.28 (a), then the difference of the CT secondary currents *i.e.* 10 $\downarrow$ 2·5 = 7·5 A will flow through the relay.

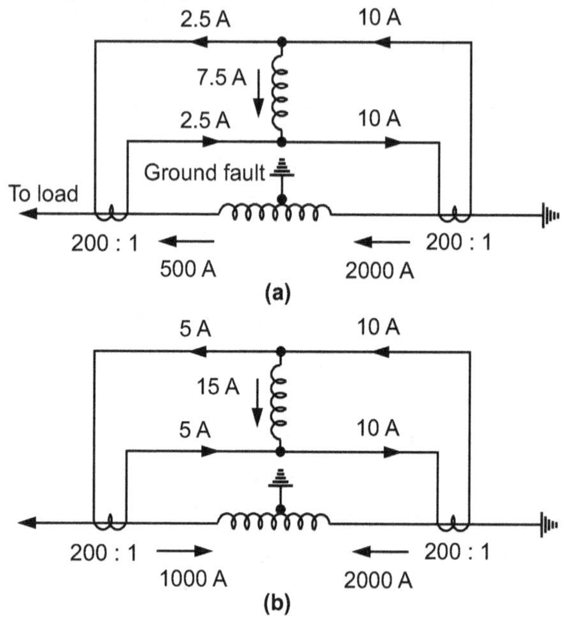

Fig. 1.28

(ii) If current flows to the fault from both sides as shown in Fig. 1.28 (b), then sum of CT secondary currents i.e. 10 + 5 = 15 A will flow through the relay.

Disadvantages

- The impedance of the pilot cables generally causes a slight difference between the currents at the two ends of the section to be protected. If the relay is very sensitive, then the small differential current flowing through the relay may cause it to operate even under no fault conditions.

- Pilot cable capacitance causes incorrect operation of the relay when a large through-current flows.

- Accurate matching of current transformers cannot be achieved due to pilot circuit impedance. The above disadvantages are overcome to a great extent in biased beam relay.

Biased Beam Relay :

- The biased beam relay (also called percentage differential relay) is designed to respond to the differential current in terms of its fractional relation to the current flowing through the protected section. Fig. 1.29 (b) shows the schematic arrangement of a biased beam relay. It is essentially an over current balanced beam relay type with an additional restraining coil. The restraining coil produces a bias force in the opposite direction to the operating force.

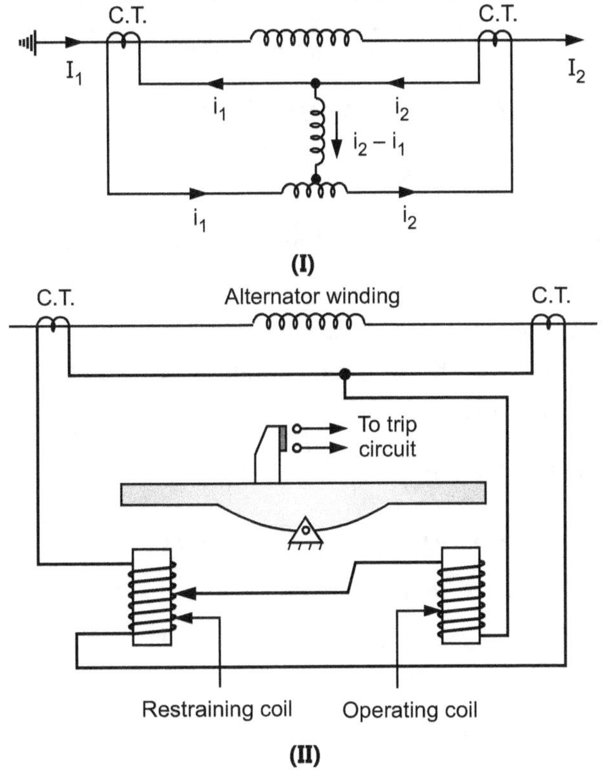

Fig. 1.29

- Under normal and through load conditions, the bias force due to restraining coil is greater than the operating force. Therefore, the relay remains inoperative. When an internal fault occurs, the operating force exceeds the bias force. Consequently, the trip contacts are closed to open the circuit breaker. The bias force can be adjusted by varying the number of turns on the restraining coil.

- The equivalent circuit of a biased beam relay is shown in Fig. 1.29 (a). The differential current in the operating coil is proportional to $i_2 - i_1$ and the equivalent current in the restraining coil is proportional to $(i_1 + i_2)/2$ since the operating coil is connected to the mid-point of the restraining coil.

- It is clear that greater the current flowing through the restraining coil, the higher the value of current required in the operating winding to trip the relay. Thus under a heavy load, a greater differential current through the relay operating coil is required for operation than under light load conditions. This relay is called percentage relay because the operating current required to trip can be expressed as a percentage of load current.

2. Voltage Balance Differential Relay

Fig. 1.30 shows the arrangement of voltage balance protection. In this scheme of protection, two similar current transformers are connected at either end of the element to be protected (e.g. an alternator winding) by means of pilot wires. The secondaries of current transformers are connected in series with a relay in such a way that under normal conditions, their induced e.m.f.s' are in opposition.

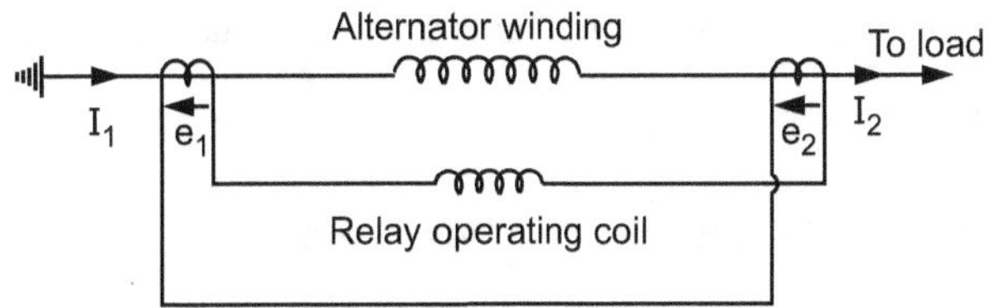

Fig. 1.30

Under healthy conditions, equal currents ($I_1 = I_2$) flow in both primary windings. Therefore, the secondary voltages of the two transformers are balanced against each other and no current will flow through the relay operating coil. When a fault occurs in the protected zone, the currents in the two primaries will differ from one another (i.e. $I_1 \neq I_2$) and their secondary voltages will no longer be in balance. This voltage difference will cause a current to flow through the operating coil of the relay which closes the trip circuit.

Disadvantages :

The voltage balance system suffers from the following drawbacks :

- A multi-gap transformer construction is required to achieve the accurate balance between current transformer pairs.

- The system is suitable for protection of cables of relatively short lengths due to the capacitance of pilot wires. On long cables, the charging current may be sufficient to operate the relay even if a perfect balance of current transformers is attained.

The above disadvantages have been overcome in Translay (modified) balanced voltage system.

1.13 PERCENTAGE DIFFERENTIAL RELAY

The simple differential relay can be made more stable, if somehow, a restraining torque proportional to the 'through fault' current could be developed-the operating torque still being proportional to the spill current. This idea has been implemented in the percentage differential relay shown in Fig. 1.31. This relay has a restraining coil which is tapped at the centre, thus forming two sections with equal number of turns, N_s. The restraining coil is connected in the circulating current path, thus receiving the 'through fault' current. The operating coil, having No number of turns, is connected in the spill path. Let us work out the torque equation for this relay.

Ampere-turns acting on the left-hand section of the restraining coil = $\dfrac{N_r}{2} I_1$.

Ampere-turns acting on the right-hand section of the restraining coil = $\dfrac{N_r}{2} I_2$

Total ampere-turns acting on the restraining coil

$$= \frac{N_r}{2} (I_1 + I_2)$$

Noting that torque in an electromagnetic relay is proportional to the square of the flux,

Torque produced by the restraining coil = $M \left[N_r \dfrac{(I_1 + I_2)}{2} \right]^2$

where M is a constant of proportionality.

Restraining torque produced by control spring = T_{spring}.

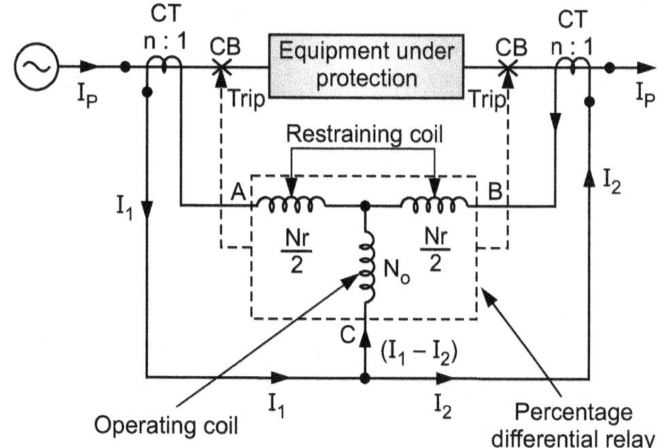

Fig. 1.31

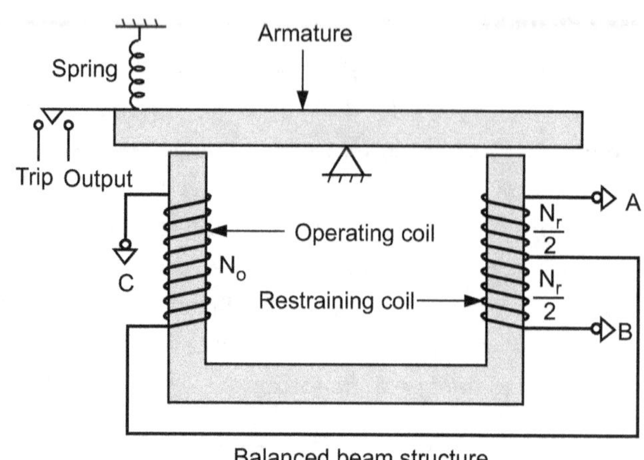

Fig. 1.32 : Percentage differential relay

Total restraining torque = $M \left[N_r \dfrac{(I_1 + I_2)}{2} \right]^2 + T_{spring}$

Similarly, Operating torque = $M [N_o (I_1 - I_2)]^2$

The relay trips if the operating torque is greater than the restraining torque. The relay will be on the verge of operation when the operating torque just balances out the restraining torque, i.e. when:

$$M [N_o (I_1 - I_2)]^2 = M \left[N_r \frac{(I_1 + I_2)}{2} \right]^2$$

(Neglecting the restraining torque due to spring)

which can be written as

$$I_1 - I_2 = K \frac{(I_1 - I_2)}{2}$$

where $K = \dfrac{N_r}{N_o}$

However, if we take into account the affect of control spring, the above equation can be written as

$$I_1 - I_2 = K \frac{(I_1 + I_2)}{2} + K_0$$

where K_0 accounts for the effect of spring.

Thus, the operating characteristics of this relay will be a straight line with a slope of (N_r/N_o) and an intercept K_0 on y-axis. All points above the straight line will represent the condition where the operating torque is greater than the restraining torque and hence will fall in the trip region of the relay. All points below the straight line belong to the restraining region. The operating characteristics of the percentage differential relay are shown in Fig. 1.33.

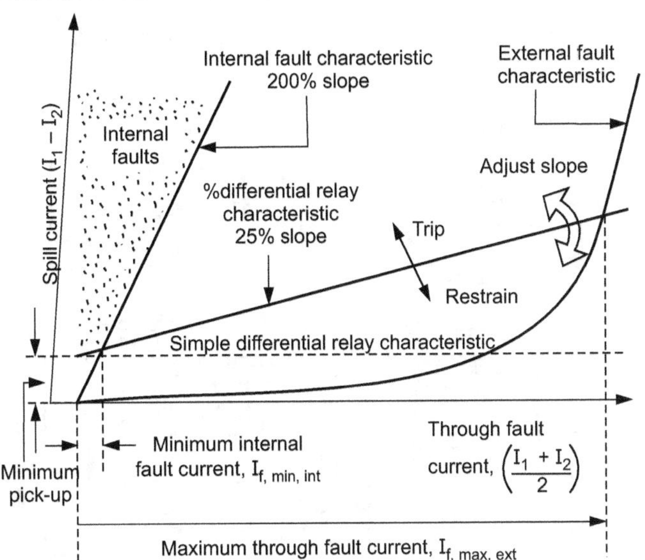

Fig. 1.33 : Operating characteristics of the percentage differential relay

Thus, the spill current must be greater than a definite percentage of the 'through fault' current for the relay to operate. Hence, the name percentage differential relay. The slope of the relay is customarily expressed as a percentage. Thus, a slope of 0.4 is expressed as 40% slope. The percentage differential relay does not have a fixed pick-up value. The relay automatically adapts its pick-up value to the 'through fault' current. As the 'through fault' current goes on increasing, we are in effect asking the relay to take it easy, by introducing a restraining torque proportional to the circulatillg current. It can be seen from Figure 1.33, that the 'through fault' stability and the stability ratio of the percentage differential relay is substantially better than that of the simple differential relay. The restraining winding is also known as the biasing winding because we bias the relay towards restraint. The slope of the characteristic is also known as *percentage* bias. The characteristic of the percentage differential relay, superimposed on the 'through fault' characteristic, and the internal fault characteristic are shown in Fig. 1.33. The slope of the internal fault characteristic can be found as follows:

Consider an internal fault in the case of a single-end-fed system. Since CT_2 will no contribute any current, i.e., $I_2 = 0$, the spill current which is $\left[\dfrac{(I_1 + I_2)}{2}\right]$ will be equal to $(I_1/2)$.

Thus, the following currents will exist during an internal fault :

$$\text{Spill current } I_1 - I_2 = I_1$$

$$\text{Circulating current } \frac{I_1 + I_2}{2} = \frac{I_1}{2}$$

Thus, during internal faults the spill current will be two times the circulating current giving a slope of 2, which is expressed as 200 %.

The minimum internal fault current below which the scheme will not respond is seen to be $I_{F, min, int}$ as shown in Fig. 1.33 Thus, the stability ratio is given by

$$\text{Stability ratio} = \frac{I_{f, max, ext}}{I_{f, min, int}}$$

The percentage differential relay can be made more immune to maloperation on 'through fault' by increasing the slope of the characteristic.

1.13.1 Block Diagram of Percentage Differential Relay

Figure 1.34 shows the block diagram of the percentage differential relay. The relay has two settings. The slope setting and the minimum pick-up setting. The slope is adjusted by changing the tapping on the restraining coil. It may be noted that both halves of the restraining coil need to be symmetrically tapped. The minimum pick-up is adjusted by changing the tension of the restraining spring.

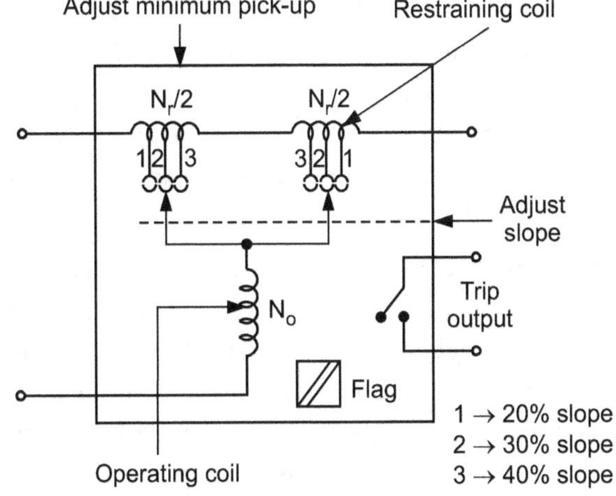

Fig. 1.34 : Block diagram of percentage differential relay showing its settings

1.13.2 Translay System

This system is the modified form of voltage-balance system. Although the principle of balanced (opposed) voltages is retained, it differs from the above voltage-balance system in that the balance or opposition is between voltages induced in the secondary coils wound on the relay magnets and *not* between the secondary voltages of the line current transformers. Since the current transformers used with Translay scheme have only to supply to a relay coil, they can be made of normal design without any air gaps. This permits the scheme to be used for feeders of any voltage.

Constructional Details :

- Fig. 1.35 shows the simplified diagram illustrating the principle of Translay scheme. It consists of two identical double winding induction type relays fitted at either end of the feeder to be protected. The primary circuits (11, 11a) of these relays are supplied through a pair of current transformers.

- The secondary windings (12, 13 and 12a, 13a) of the two relays are connected in series by pilot wires in such a way that voltages induced in the former opposes the other. The compensating devices (18, 18a) neutralise the effects of pilot-wire capacitance currents and of inherent lack of balance between the two current transformers.

Operation :

- Under healthy conditions, current at the two ends of the protected feeder is the same and the primary windings (11, 11a) of the relays carry the same current. The windings 11 and 11a induce equal e.m.f.s in the secondary windings 12, 12a and 13, 13a. As these windings are so connected that their induced voltages are in opposition, no current will flow through the pilots or operating coils and hence no torque will be exerted on the disc of either relay.

- In the event of fault on the protected feeder, current leaving the feeder will differ from the current entering the feeder. Consequently, unequal voltages will be induced in the secondary windings of the relays and current will circulate between the two windings, causing the torque to be exerted on the disc of each relay.

- As the direction of secondary current will be opposite in the two relays, therefore, the torque in one relay will tend to close the trip circuit while in the other relay, the torque will hold the movement in the normal unoperated position. It may be noted that resulting operating torque depends upon the position and nature of the fault in the protected zone and atleast one element of either relay will operate under any fault condition.

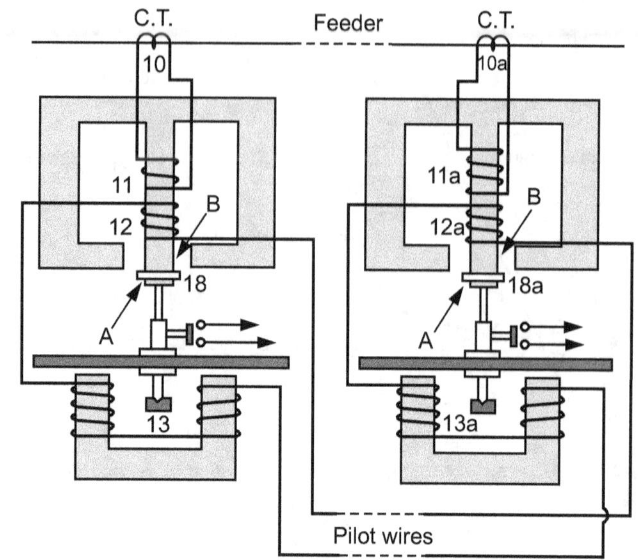

Fig. 1.35

- It is worthwhile here to mention the role of closed copper rings (18, 18a) in neutralising the effects of pilot capacitive currents. Capacitive currents lead the voltage impressed across the pilots by 90° and when they flow in the operating winding 13 and 13a (which are of low inductance), they produce fluxes that also lead the pilot voltage by 90°.

- Since pilot voltage is that induced in the secondary windings 12 and 12a, it lags by a substantial angle behind the fluxes in the field magnet air gaps *A* and *B*. The closed copper rings (18, 18a) are so adjusted that this angle is approximately 90°. In this way fluxes acting on the disc are in phase and hence no torque is exerted on the relay disc.

1.14 STATIC RELAYS

- In a static relay, the comparison or measurement of electrical quantities is performed by a static circuit which gives an output signal for the tripping of a circuit breaker. Most of the present day static relays include a dc polarised relay as a slave relay. The slave relay is an output device and does not perform the function of comparison or measurement. It simply closes contacts. It is used because of its low cost. In a fully static relay, a thyristor is used in place of the electromagnetic slave relay. The electromechanical relay used as a slave relay provides a number of output contacts at low cost. Electromagnetic multicontact tripping arrangements are much simpler than an equivalent group of thyristor circuits.

- A static relay (or solid state relay) employs semiconductor diodes, transistors, zener diodes, thyristors, logic gates, etc. as its components. Now-a-days, integrated circuits are being used in place of transistors. They are more reliable and compact.

- Earlier, induction cup units were widely used for distance and directional relays. Later these were replaced by rectifier bridge type static relays which employed dc polarised relays as slave relays. Where overcurrent relays are needed, induction disc relays are in universal use throughout the world. But ultimately static relays will supersede all electromagnetic relays, except the attracted armature relays and depolarised relays as these relays can control many circuit at low costs.

Merits and Demerits of Static Relays

The advantages of static relays over electromechanical relays are as follows.

- Low burden on CTs and PTs. The static relays consume less power and in most of the cases they draw power from the auxiliary dc supply.

- Fast response

- Long life

- High resistance to shock and vibration

- Less maintenance due to the absence of moving parts and bearings

- Frequent operations cause no deterioration

- Quick resetting and absence of overshoot

- Compact size

- Greater sensitivity as amplification can be provided easily

- Complex relaying characteristics can easily be obtained

- Logic circuits can be used for complex protective schemes

The logic circuit may take decisions to operate under certain conditions and not to operate under other conditions.

The demerits of static relays are as follows:

- Static relays are temperature sensitive. Their characteristics may vary with the variation of temperature. Temperature compensation can be made by using thermistors and by using digital techniques for measurements, etc.

- Static relays are sensitive to voltage transients. The semiconductor components may get damaged due to voltage spikes. Filters and shielding can be used for their protection against voltage spikes.

- Static relays need an auxiliary power supply. This can however be easily supplied by a battery or a stabilized power supply.

1.14.1 Comparators

- When faults occur on a system, the magnitude of voltage and current and phase angle between voltage and current may change. These quantities during faulty conditions are different from those under healthy conditions. The static relay circuitry is designed to recognise the changes and to distinguish between healthy and faulty conditions.

- Either magnitudes of voltage/current (or corresponding derived quantities) are compared or phase angle between voltage and current (or corresponding derived quantities) are measured by the static relay circuitry and a trip signal is sent to the circuit breaker when a fault occurs. The part of the circuitry which compares the two actuating quantities either in amplitude or phase is known as the comparator. There are two types of comparators—amplitude comparator and phase comparator.

Amplitude Comparator

- An amplitude comparator compares the magnitudes of two input quantities, irrespective of the angle between them. One of the input quantities is an operating quantity and the other a restraining quantity. When the amplitude of the operating quantity exceeds the amplitude of the restraining quantity, the relay sends a tripping signal.

Phase Comparator

- A phase comparator compares two input quantities in phase angle, irrespective of their magnitudes and operates if the phase angle between them is < 90°.

Duality between Amplitude and Phase Comparators

- An amplitude comparator can be converted to a phase comparator and vice versa if the input quantities to the comparator are modified. The modified input quantities arc the sum and difference of the original two input quantities. To understand this fact, consider the operation of an amplitude comparator which has two input signals M and was shown in Fig. 1.36(a).

- It operates when (M) > (N). Now change the input quantities to (M + N) and (M − N) as shown in Fig. 1.36(b). As its circuit is designed for amplitude comparison, now with the changed input, it will operate when (M + N) > (M − N). This condition will be satisfied only when the phase angle between M and N is less than 90°. This has been illustrated with the phasor diagram shown in Fig. 1.37. It means that the comparator with the modified inputs has now become a phase comparator for the original input signals H and N.

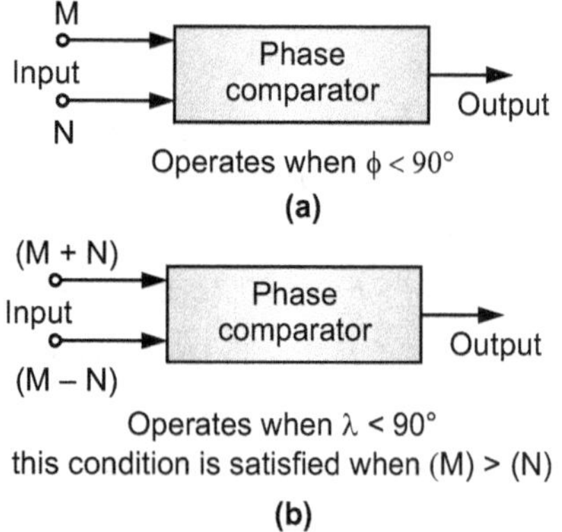

Fig. 1.36: (a) Amplitude comparator (b) Amplitude comparator used for phase comparison

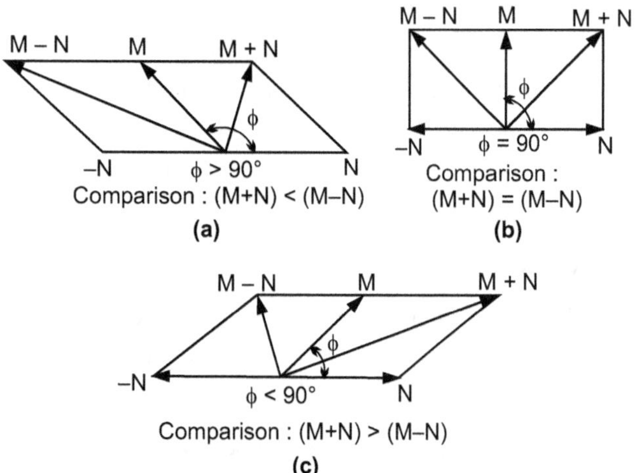

Fig. 1.37: Phasor diagram for amplitude comparator used for phase comparison

Similarly, consider a phase comparator shown in Fig. 1.37 (a). It compares the phases of input signals M and N. If the phase angle between M and N, i.e. angle φ is less than 90°, the comparator operates. Now change the input, signals to (M + N) and (M − N), as in Fig.1.37 (b) . With these changed inputs the comparator will operate when phase angle between (M + N) and (M − N), i.e. angle λ is less than 90°. This condition will be satisfied only when (M) > (N). In other words, the phase comparator with changed inputs has now become an amplitude comparator for the original input signals M and N. This has been illustrated with phasor diagrams as shown in Fig.1.38

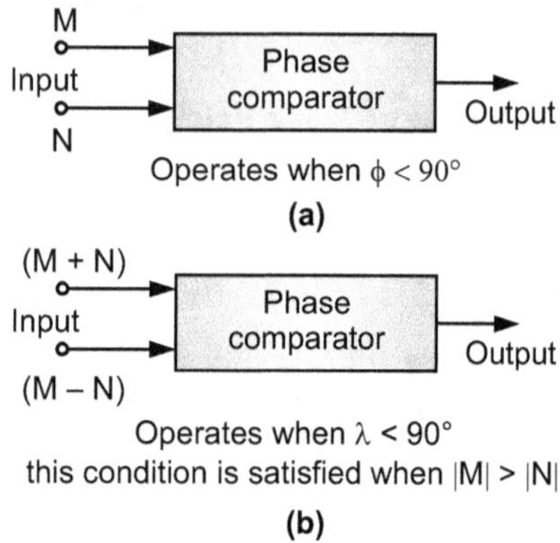

Fig. 1.38 :(a) Phase comparator (b) Phase comparator used for amplitude comparison

Fig. 1.38 shows three phasor diagrams for an amplitude comparator. The phase angle between the original inputs M and N is 0. Now the inputs to the amplitude comparator are changed to (M + N) and (M − N) and its behaviour is examined with the help of three phasor diagrams. The three phasor diagrams are with phase angle 0

(i)   greater than 90°,

(ii) equal to 90° and

(iii) less than 90°, respectively.

When 0 is less than 90°, (M + N) become greater than (M − N) and the relay operates with the modified inputs. When 0 is equal to 90° or greater than 90°, the relay does not operate.

The phasor diagrams show that IM + NI becomes greater than (M − N) only when 0 is less than 90°. This will be true irrespective of the magnitude of M and N. In other words, this will be true whether (M) = N) or (M) > (N) or (M) < (N). The Fig. have been drawn with (M) = (N). The reader can draw phasor diagrams with (M) < (M) or (M) > (N). The results will remain the same. This shows that with changed inputs, the amplitude comparator is converted to a phase comparator for the original inputs.

Fig. 1.38 (b) shows three phasor diagrams for a phase comparator. The original inputs are M and N. Now the inputs of the phase comparators are changed to (M + N) and (M - N), and its behavior is examined with the help of three phasor diagrams drawn for (i) (M) < (N), (ii) (M) = (N) and (iii) (M) > (N). The angle between (M + N) and (M - N) is A. The angle A becomes less than 90° only when (M) > (N). As the comparator under consideration is a phase comparator, the relay will trip. But for the original inputs M and N, the comparator behaves as an amplitude comparator. This will be true irrespective of the phase angle ϕ between M and N. The figure has been drawn with ϕ less than 90°. The reader can check it by drawing phasors with $\phi = 90°$ or $\phi > 90°$. The result will remain the same.

1.15 TYPES OF PROTECTION

When a fault occurs on any part of electric power system, it must be cleared quickly in order to avoid damage and/or interference with the rest of the system. It is a usual practice to divide the protection scheme into two classes *viz.* primary protection and back-up protection.

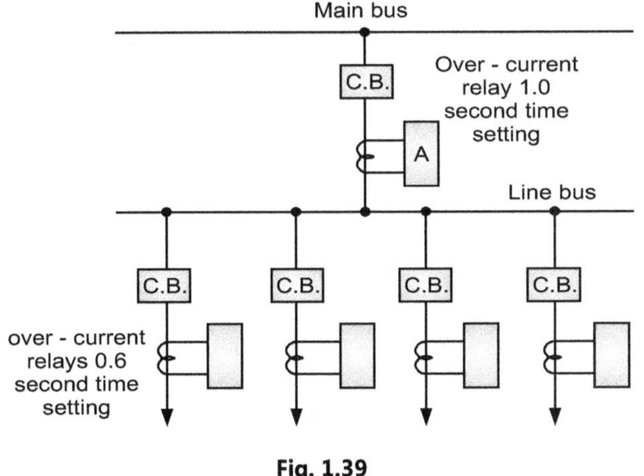

Fig. 1.39

1. **Primary Protection :** It is the protection scheme which is designed to protect the component parts of the power system. Thus referring to Fig. 1.39, each line has

an overcurrent relay that protects the line. If a fault occurs on any line, it will be cleared by its relay and circuit breaker. This forms the primary or main protection and serves as the first line of defence. The service record of primary relaying is very high with well over ninety percent of all operations being correct. However, sometimes faults are not cleared by primary relay system because of trouble within the relay, wiring system or breaker. Under such conditions, back-up protection does the required job.

2. **Back-up Protection :**

- It is the second line of defence in case of failure of the primary protection. It is designed to operate with sufficient time delay so that primary relaying will be given enough time to function if it is able to. Thus referring to Fig. 1.39, relay A provides back-up protection for each of the four lines.

- If a line fault is not cleared by its relay and breaker, the relay A on the group breaker will operate after a definite time delay and clear the entire group of lines. It is evident that when back-up relaying functions, a larger part is disconnected than when primary relaying functions correctly. Therefore, greater emphasis should be placed on the better maintenance of primary relaying.

EXERCISE

1. Describe the various steps for calculating the actual relay operating time.

2. Explain with the help of neat diagram the construction and working of :

    (i) Non-directional induction type overcurrent relay

    (ii) Induction type directional power relay

    (iii) Describe the construction and principle of operation of an induction type directional overcurrent relay.

3. Explain the working principle of distance relays.

4. Write a detailed note on differential relays.

5. Describe the Translay scheme of protection.

6. Write short notes on the following.

    (i)   Percentage differential relays

    (ii) Definite distance relays

    (iii) Time-distance relays.

7. What is protective relay ? Explain its function in an electrical system.

8. Discuss the fundamental requirements of protective relaying.

9. Describe briefly some important types of electromagnetic attraction relays.

10. Derive the equation for torque developed in an induction relay.

11. Write a brief note on relay timing.

12. Explain static relay and its merits and demerits.

PRINCIPLES OF CIRCUIT BREAKERS

2.1 INTRODUCTION

- During the operation of power system, it is often desirable and necessary to switch on or off the various circuits (*e.g.*, transmission lines, distributors, generating plants etc.) under both normal and abnormal conditions. In earlier days, this function used to be performed by a switch and a fuse placed in series with the circuit. However, such a means of control presents two disadvantages.

- Firstly, when a fuse blows out, it takes quite sometime to replace it and restore supply to the customers. Secondly, a fuse cannot successfully interrupt heavy fault currents that result from faults on modern high-voltage and large capacity circuits. Due to these disadvantages, the use of switches and fuses is limited to low-voltage and small capacity circuits where frequent operations are not expected *e.g.*, for switching and protection of distribution transformers, lighting circuits, branch circuits of distribution lines etc.

- With the advancement of power system, the lines and other equipment operate at very high voltages and carry large currents. The arrangement of switches along with fuses cannot serve the desired function of swithgear in such high capacity circuits. This necessitates to employ a more dependable means of control such as is obtained by the use of *circuit breakers*. A circuit breaker can make or break a circuit either manually or automatically under all conditions *viz.*, no-load, full-load and short-circuit conditions. This characteristic of the circuit breaker has made it a very useful equipment for switching and protection of various parts of the power system. In this chapter, we shall deal with the various types of circuit breakers and their increasing applications as control devices.

2.2 CIRCUIT BREAKERS

A circuit breaker is a piece of equipment which can :

(i) Make or break a circuit either manually or by remote control under normal conditions

(ii) Break a circuit automatically under fault conditions

(iii) Make a circuit either manually or by remote control under fault conditions

Thus a circuit breaker incorporates manual (or remote control) as well as automatic control for switching functions. The latter control employs relays and operates only under fault conditions.

Operating Principle :

- A circuit breaker essentially consists of fixed and moving contacts, called electrodes. Under normal operating conditions, these contacts remain closed and will not open automatically until and unless the system becomes faulty. Of course, the contacts can be opened manually or by remote control whenever desired. When a fault occurs on any part of the system, the trip coils of the circuit breaker get energised and the moving contacts are pulled apart by some mechanism, thus opening the circuit.

- When the contacts of a circuit breaker are separated under fault conditions, an arc is struck between them. The current is thus able to continue until the discharge ceases. The production of arc not only delays the current interruption process but it also generates enormous heat which may cause damage to the system or to the circuit breaker itself. Therefore, the main problem in a circuit breaker is to extinguish the arc within the shortest possible time so that heat generated by it may not reach a dangerous value.

2.2.1 D.C Circuit Breakers

DC circuit breaker, like their name suggests, is used for the protection of electrical devices that operate with direct current. The main difference between direct current and alternating current is that in DC the voltage output is constant, while in AC it cycles several times per second.

Miniature circuit breakers available for use in direct current Nowadays we use more commonly miniature circuit breaker or MCB in low voltage electrical network instead of fuse. The MCB has some advantages compared to fuse :

- It automatically switches off the electrical circuit during abnormal condition of the network means in over load condition as well as faulty condition. The fuse does not sense but miniature circuit breaker does it in more reliable way. MCB is much more sensitive to over current than fuse.

- Another advantage is, as the switch operating knob comes at its off position during tripping, the faulty zone of the electrical circuit can easily be identified. But in case of fuse, fuse wire should be checked by opening fuse grip or cutout from fuse base, for confirming the blow of fuse wire.

- Quick restoration of supply cannot be possible in case of fuse as because fuses have to be replaced for restoring the supply. But in the case of MCB, quick restoration is possible by just switching on operation.

- Handling MCB is more electrically safe than fuse. Because of too many advantages of MCB over fuse units, in modern low voltage electrical network, miniature circuit breaker is mostly used instead of backdated fuse unit. Only one disadvantage of MCB over fuse is that this system is more costly than fuse unit system.

Working Principle Miniature Circuit Breaker

- There are two arrangement of operation of miniature circuit breaker. One due to thermal effect of over current and other due to electromagnetic effect of over current. The thermal operation of miniature circuit breaker is achieved with a bimetallic strip whenever continuous over current flows through MCB, the bimetallic strip is heated and deflects by bending. This deflection of bimetallic strip releases mechanical latch.

- As this mechanical latch is attached with operating mechanism, it causes to open the miniaturecircuit breaker contacts. But during short circuit condition, sudden rising of current, causes electromechanical displacement of plunger associated with tripping coil or solenoid of MCB. The plunger strikes the trip lever causing immediate release of latch mechanism consequently open the circuit breaker contacts.

- This was a simple explanation of miniature circuit breaker working principle.

Miniature Circuit Breaker Construction

- Miniature circuit breaker construction is very simple, robust and maintenance free. Generally a MCB is not repaired or maintained, it just replaced by new one when required. A miniature circuit breaker has normally main constructional parts. These are:
  - ➢ Frame of Miniature Circuit Breaker
  - ➢ The frame of miniature circuit breaker is a molded case. This is a rigid, strong, insulated housing in which the other components are mounted.

Operating Mechanism of Miniature Circuit Breaker

- The operating mechanism of miniature circuit breaker provides the means of manual opening and closing operation of miniature circuit breaker. It has three-positions "ON," "OFF," and "TRIPPED". The external switching latch can be in the "TRIPPED" position, if the MCB is tripped due to over-current. When manually switch off the MCB, the switching latch will be in "OFF" position. In close condition of MCB, the switch is positioned at "ON". By observing the positions of the switching latch one can determine the condition of MCB whether it is closed, tripped or manually switched off.

Trip Unit of Miniature Circuit Breaker

- The trip unit is the main part, responsible for proper working of miniature circuit breaker. Two main types of trip mechanism are provided in MCB. A bimetal provides protection against over load current and an electromagnet provides protection against shortcircuit current.

Operation of Miniature Circuit Breaker

- There are three mechanisms provided in a single miniature circuit breaker to make it switched off. If we carefully observe the picture beside, we will find there are mainly one bi – metallic strip, one trip coil and one hand operated on – off lever. Electric current carrying path of a miniature circuit breaker shown in the picture is like follows. First left hand side power terminal – then bimetallic strip – then current coil or trip coil – then moving contact – then fixed contact and – lastly right had side power terminal. All are arranged in series.

- If circuit is overloaded for long time, the bi – metallic strip becomes over heated and deformed. This deformation of bi metallic strip causes, displacement of latch point. The moving contact of the MCB is so arranged by means of spring pressure, with this latch point, that a little displacement of latch causes, release of spring and makes the moving contact to move for opening the MCB.

- The current coil or trip coil is placed such a manner, that during short circuit fault the mmf of that coil causes its plunger to hit the same latch point and make the latch to be displaced. Hence the MCB will open in same manner. Again when operating lever of the miniature circuit breaker is operated by hand, that means when we make the MCB at off position manually, the same latch point is displaced as a result moving contact separated from fixed contact in same manner.

- So, whatever may be the operating mechanism, that means, may be due to deformation of bi – metallic strip, due to increased mmf of trip coil or may due to manual operation, actually the same latch point is displaced and same deformed spring is released, which ultimately responsible for movement of the moving contact. When the the moving contact separated from fixed contact, there may be a high chance of arc.

2.2.2 A.C Circuit Breaking

- There is difference between breaking in case of d.c and a.c circuits. In a.c circuits the current passes through zero twice in one complete cycle. When the currents are reduced to zero the breakers are operated to cut off the current. This will avoid the striking of the arc. But this conditions is difficult to achieve and very much expensive.

- The restriking of arc when current is interrupted is dependent on the voltage between the contact gap at that instant which will inturn depend on power factor. Higher the factor ,lesser is the voltage appearing across the gap than its peak value.

2.3 ARC PHENOMENON

- When a short-circuit occurs, a heavy current flows through the contacts of the circuit breaker before they are opened by the protective system. At the instant when the contacts begin to separate, the contact area decreases rapidly and large fault current causes increased current density and hence rise in temperature.

- The heat produced in the medium between contacts (usually the medium is oil or air) is sufficient to ionise the air or vapourise and ionise the oil. The ionised air or vapour acts as conductor and an arc is struck between the contacts. The p.d. between the contacts is quite small and is just sufficient to maintain the arc.

- The arc provides a low resistance path and consequently the current in the circuit remains uninterrupted so long as the arc persists. During the arcing period, the current flowing between the contacts depends upon the arc resistance. The greater the arc resistance, the smaller the current that flows between the contacts.

The arc resistance depends upon the following factors:

(i) Degree of Ionization : The arc resistance increases with the decrease in the number of ionized particles between the contacts.

(ii) Length of the Arc : the arc resistance increases with the length of the arc *i.e.,* separation of contacts.

(iii) Cross-section of Arc : the arc resistance increases with the decrease in area of X-section of the arc.

2.4 PRINCIPLES OF ARC EXTINCTION

Before discussing the methods of arc extinction, it is necessary to examine the factors responsible for the maintenance of arc between the contacts. These are :

(i) P.D. between the contacts

(ii) Ionised particles between contacts

Taking these in turn,

(i) When the contacts have a small separation, the P.D. between them is sufficient to maintain the arc. One way to extinguish the arc is to separate the contacts to such a distance that P.D. becomes inadequate to maintain the arc. However, this method is impracticable in high voltage system where a separation of many metres may be required.

(ii) The ionised particles between the contacts tend to maintain the arc. If the arc path is deionised, the arc extinction will be facilitated. This may be achieved by cooling the arc or by bodily removing the ionised particles from the space between the contacts.

2.5 METHODS OF ARC EXTINCTION

There are two methods of extinguishing the arc in circuit breakers *viz.*

    1. High resistance method.

    2. Low resistance or current zero method

1. High Resistance Method :

- In this method, arc resistance is made to increase with time so that current is reduced to a value insufficient to maintain the arc. Consequently, the current is interrupted or the arc is extinguished. The principal disadvantage of this method is that enormous energy is dissipated in the arc. Therefore, it is employed only in d.c. circuit breakers and low-capacity a.c. circuit breakers. The resistance of the arc may be increased by :

  ➢ **Lengthening the Arc :** The resistance of the arc is directly proportional to its length. The length of the arc can be increased by increasing the gap between contacts.

  ➢ **Cooling the Arc :** Cooling helps in the deionisation of the medium between the contacts. This increases the arc resistance. Efficient cooling may be obtained by a gas blast directed along the arc.

- ➢ **Reducing X-section of the Arc :** If the area of X-section of the arc is reduced, the voltage necessary to maintain the arc is increased. In other words, the resistance of the arc path is increased. The cross-section of the arc can be reduced by letting the arc pass through a narrow opening or by having smaller area of contacts.

- ➢ **Splitting the Arc :** The resistance of the arc can be increased by splitting the arc into a number of smaller arcs in series. Each one of these arcs experiences the effect of lengthening and cooling. The arc may be split by introducing some conducting plates between the contacts.

2. **Low Resistance or Current Zero Method :**

- This method is employed for arc extinction in a.c. circuits only. In this method, arc resistance is kept low until current is zero where the arc extinguishes naturally and is prevented from restriking inspite of the rising voltage across the contacts. All modern high power a.c. circuit breakers employ this method for arc extinction.

- In an a.c. system, current drops to zero after every half-cycle. At every current zero, the arc extinguishes for a brief moment. Now the medium between the contacts contains ions and electrons so that it has small dielectric strength and can be easily broken down by the rising contact voltage known as restriking voltage. If such a breakdown does occur, the arc will persist for another halfcycle.

- If immediately after current zero, the dielectric strength of the medium between contacts is built up more rapidly than the voltage across the contacts, the arc fails to restrike and the current will be interrupted. The rapid increase of dielectric strength of the medium near current zero can be achieved by :

(a) Causing the ionised particles in the space between contacts to recombine into neutral molecules.

(b) Sweeping the ionised particles away and replacing them by un-ionised particles

- Therefore, the real problem in a.c. arc interruption is to rapidly deionise the medium between contacts as soon as the current becomes zero so that the rising contact voltage or restriking voltage cannot breakdown the space between contacts. The de-ionisation of the medium can be achieved by:

- ➢ **Lengthening of the Gap :** The dielectric strength of the medium is proportional to the length of the gap between contacts. Therefore, by opening the contacts rapidly, higher dielectric strength of the medium can be achieved.

- ➢ **High Pressure :** If the pressure in the vicinity of the arc is increased, the density of the particles constituting the discharge also increases. The increased density of particles causes higher rate of de-ionisation and consequently the dielectric strength of the medium between contacts is increased.

- ➢ **Cooling :** Natural combination of ionised particles takes place more rapidly if they are allowed to cool. Therefore, dielectric strength of the medium between the contacts can be increased by cooling the arc.

- ➢ **Blast Effect :** If the ionised particles between the contacts are swept away and replaced by unionized particles, the dielectric strength of the medium can be increased considerably. This may be achieved by a gas blast directed along the discharge or by forcing oil into the contact space.

2.6 IMPORTANT TERMS

The following are the important terms much used in the circuit breaker analysis :

(i) Arc Voltage :

- It is the voltage that appears across the contacts of the circuit breaker during the arcing period.

- As soon as the contacts of the circuit breaker separate, an arc is formed. The voltage that appears across the contacts during arcing period is called the arc voltage.

- Its value is low except for the period the fault current is at or near zero current point. At current zero, the arc voltage rises rapidly to peak value and this peak voltage tends to maintain the current flow in the form of arc.

(ii) Restriking Voltage :

- It is the transient voltage that appears across the contacts at or near current zero during arcing period. At current zero, a high-frequency transient voltage appears across the contacts and is caused by the rapid distribution of energy between the magnetic and electric fields associated with the plant and transmission lines of the system. This transient voltage is known as restriking voltage (Fig. 2.1).

- The current interruption in the circuit depends upon this voltage. If the restriking voltage rises more rapidly than the dielectric strength of the medium between the contacts, the arc will persist for another half-cycle.

On the other hand, if the dielectric strength of the medium builds up more rapidly than the restriking voltage, the arc fails to restrike and the current will be interrupted.

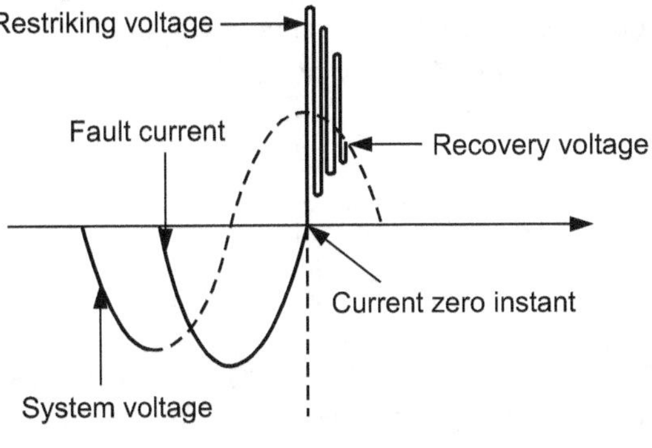

Fig. 2.1

(iii) Recovery Voltage :

- It is the normal frequency (50 Hz) R.M.S. voltage that appears across the contacts of the circuit breaker after final arc extinction. It is approximately equal to the system voltage. When contacts of circuit breaker are opened, current drops to zero after every half cycle. At some current zero, the contacts are separated sufficiently apart and dielectric strength of the medium between the contacts attains a high value due to the removal of ionised particles.

- At such an instant, the medium between the contacts is strong enough to prevent the breakdown by the restriking voltage. Consequently, the final arc extinction takes place and circuit current is interrupted. Immediately after final current interruption, the voltage that appears across the contacts has a transient part (See Fig. 2.1). However, these transient oscillations subside rapidly due to the damping effect of system resistance and normal circuit voltage appears across the contacts. The voltage across the contacts is of normal frequency and is known as recovery voltage.

2.6.1 Problems of Circuit Interruption

- The power system contains an appreciable amount of inductance and some capacitance. When a fault occurs, the energy stored in the system can be considerable. Interruption of fault current by a circuit breaker will result in most of the stored energy dissipated within the circuit breaker, the remainder being dissipated during oscillatory surges in the system. The oscillatory surges are undesirable and, therefore, the circuit

breaker must be designed to dissipate as much of the stored energy as possible.

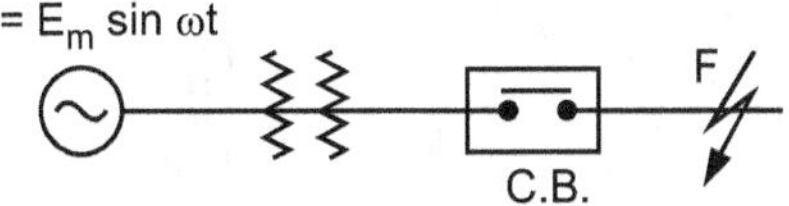

(a)

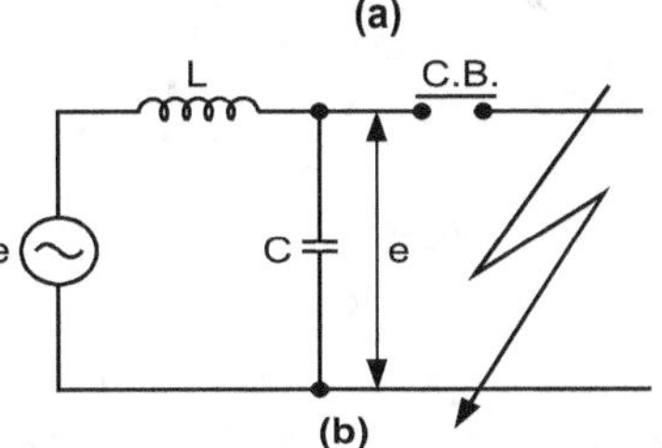

(b)

Fig. 2.2

- Fig. 2.2 (a) shows a short-circuit occurring on the transmission line. Fig 2.2 (b) shows its equivalent circuit where L is the inductance per phase of the system up to the point of fault and C is the capacitance per phase of the system. The resistance of the system is neglected as it is generally small.

(i) Rate of Rise of Re-Striking Voltage :

- It is the rate of increase of re-striking voltage and is abbreviated by R.R.R.V. Usually, the voltage is in kV and time in microseconds so that R.R.R.V. is in kV/μ sec. Consider the opening of a circuit breaker under fault conditions shown in simplified form in Fig. 2.2 (b) above. Before current interruption, the capacitance C is short-circuited by the fault and the short-circuit current through the breaker is limited by inductance L of the system only.

- Consequently, the short-circuit current will lag the voltage by 90° as shown in Fig. 2.3, where I represents the short-circuit current and e_a represents the arc voltage. It may be seen that in this condition, the entire generator voltage appears across inductance L. When the contacts are opened and the arc finally extinguishes at some current zero, the generator voltage e is suddenly applied to the inductance and capacitance in series. This L–C combination forms an oscillatory circuit and produces a transient of frequency.

$$f_n = \frac{1}{2\pi\sqrt{LC}}$$

which appears across the capacitor C and hence across the contacts of the circuit breaker.

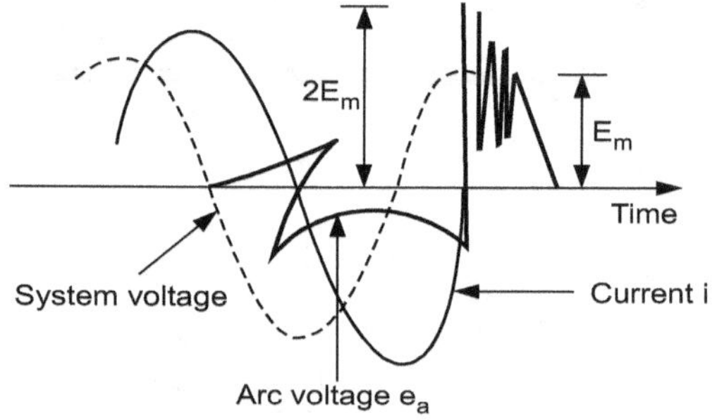

Fig. 2.3

- This transient voltage, as already noted, is known as re-striking voltage and may reach an instantaneous peak value twice the peak phase-neutral voltage i.e. $2E_m$. The system losses cause the oscillations to decay fairly rapidly but the initial overshoot increases the possibility of re-striking the arc.

- It is the rate of rise of re-striking voltage (R.R.R.V.) which decides whether the arc will re-strike or not. If R.R.R.V. is greater than the rate of rise of dielectric strength between the contacts, the arc will re-strike. However, the arc will fail to re-strike if R.R.R.V. is less than the rate of increase of dielectric strength between the contacts of the breaker. The value of R.R.R.V. depends upon :

(a) Recovery voltage

(b) Natural frequency of oscillations

- For a short-circuit occuring near the power station bus-bars, C being small, the natural frequency ($f_n = 1/2\pi LC$) will be high. Consequently, R.R.R.V. will attain a large value. Thus the worst condition for a circuit breaker would be that when the fault takes place near the bus-bars.

(ii) Current Chopping :

- It is the phenomenon of current interruption before the natural current zero is reached. Current chopping mainly occurs in air-blast circuit breakers because they retain the same extinguishing power irrespective of the magnitude of the current to be interrupted.

- When breaking low currents (e.g., transformer magnetising current) with such breakers, the powerful De-ionising effect of air-blast causes the current to fall abruptly to zero well before the natural current zero is reached. This phenomenon is known as current chopping and results in the production of high voltage transient across the contacts of the circuit breaker as discussed below:

- Consider again Fig. 2.2 (b) repeated as Fig. 2.4 (i). Suppose the arc current is i when it is chopped down to zero value as shown by point a in Fig. 2.4 (ii). As the chop occurs at current i, therefore, the energy stored in inductance is L i^2/2. This energy will be transferred to the capacitance C, charging the latter to a prospective voltage e given by :

$$\frac{1}{2}Li^2 = \frac{Ce^2}{2}$$

or $$e = i\sqrt{\frac{L}{C}}\text{ volts}$$

- The prospective voltage e is very high as compared to the dielectric strength gained by the gap so that the breaker restrikes. As the de-ionising force is still in action, therefore, chop occurs again but the arc current this time is smaller than the previous case. This induces a lower prospective voltage to re-ignite the arc. In fact, several chops may occur until a low enough current is interrupted which produces insufficient induced voltage to re-strike across the breaker gap. Consequently, the final interruption of current takes place.

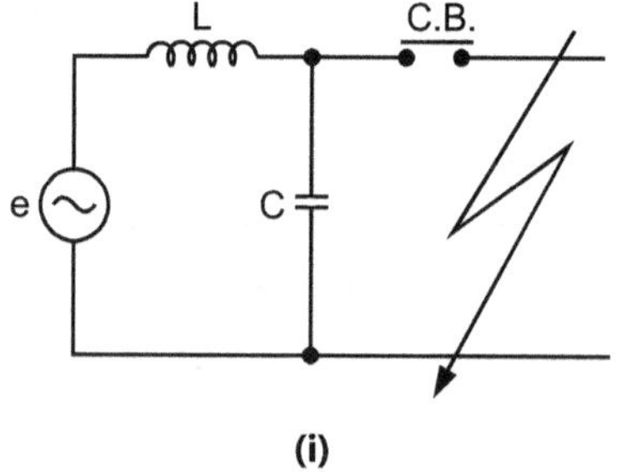

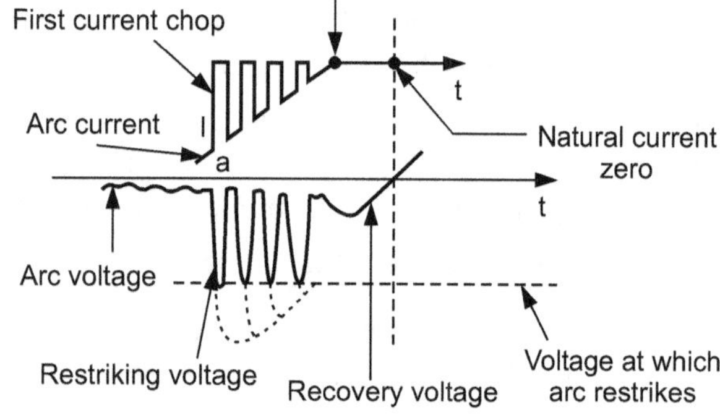

Fig. 2.4

- Excessive voltage surges due to current chopping are prevented by shunting the contacts of the breaker with a resistor (resistance switching) such that reignition is unlikely to occur.

(iii) Capacitive Current Breaking :

- Another cause of excessive voltage surges in the circuit breakers is the interruption of capacitive currents. Examples of such instances are opening of an unloaded long transmission line, disconnecting a capacitor bank used for power factor improvement etc. Consider the simple equivalent circuit of an unloaded transmission line shown in Fig. 2.5. Such a line, although unloaded in the normal sense, will actually carry a capacitive current I on account of appreciable amount of capacitance C between the line and the earth.

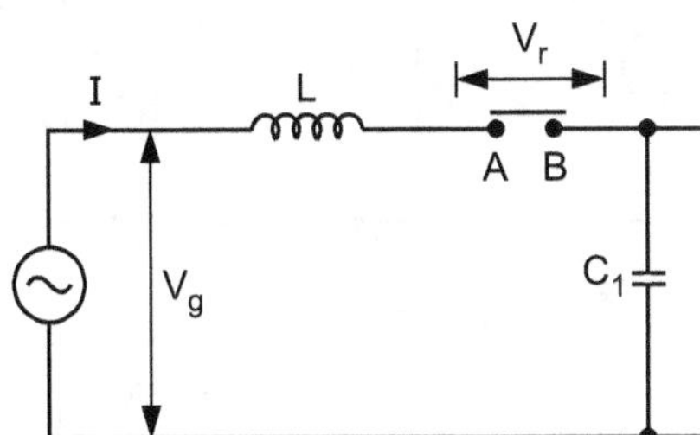

Fig. 2.5

- Let us suppose that the line is opened by the circuit breaker at the instant when line capacitive current is zero [point 1 in Fig. 2.6]. At this instant, the generator voltage V_g will be maximum (i.e., V_{gm}) lagging behind the current by 90°. The opening of the line leaves a standing charge on it (i.e., end B of the line) and the capacitor C_1 is charged to V_{gm}. However, the generator end of the line (i.e., end A of the line) continues its normal sinusoidal variations. The voltage V_r across the circuit breaker will be the difference between the voltages on the respective sides.

- Its initial value is zero (point 1) and increases slowly in the beginning. But half a cycle later [point R in Fig. 2.6], the potential of the circuit breaker contact 'A' becomes maximum negative which causes the voltage across the breaker (Vr) to become $2\,V_{gm}$. This voltage may be sufficient to restrike the arc. The two previously separated parts of the circuit will now be joined by an arc of very low resistance. The line capacitance discharges at once to reduce the voltage across the circuit breaker, thus setting up high frequency transient. The peak value of the initial transient will be twice the voltage at that instant i.e., $-4\,V_{gm}$. This will cause the transmission voltage to swing to $-4\,V_{gm}$ to $+V_{gm}$ i.e., $-3\,V_{gm}$.

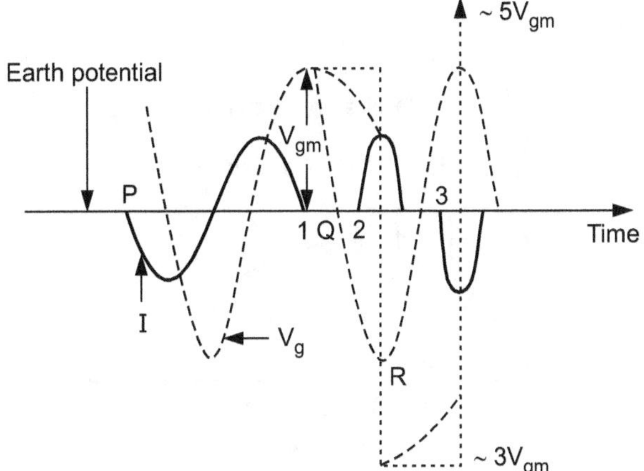

Fig. 2.6

- The re-strike arc current quickly reaches its first zero as it varies at natural frequency. The voltage on the line is now $-3\,V_{gm}$ and once again the two halves of the circuit are separated and the line is isolated at this potential. After about half a cycle further, the aforesaid events are repeated even on more formidable scale and the line may be left with a potential of $5\,V_{gm}$ above earth potential.

- Theoretically, this phenomenon may proceed infinitely increasing the voltage by successive increment of 2 times V_{gm}. While the above description relates to the worst possible conditions, it is obvious that if the gap breakdown strength does not increase rapidly enough, successive re-strikes can build up a dangerous voltage in the open circuit line. However, due to leakage and corona loss, the maximum voltage on the line in such cases is limited to $5\,V_{gm}$.

Resistance Switching

- It has been discussed above that current chopping, capacitive current breaking etc. give rise to severe voltage oscillations. These excessive voltage surges during circuit interruption can be prevented by the use of shunt resistance R connected across the circuit breaker contacts as shown in the equivalent circuit in Fig. 2.7. This is known as resistance switching.

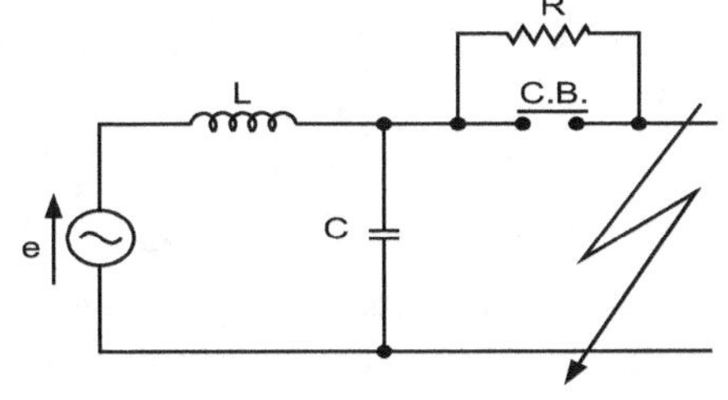

Fig. 2.7

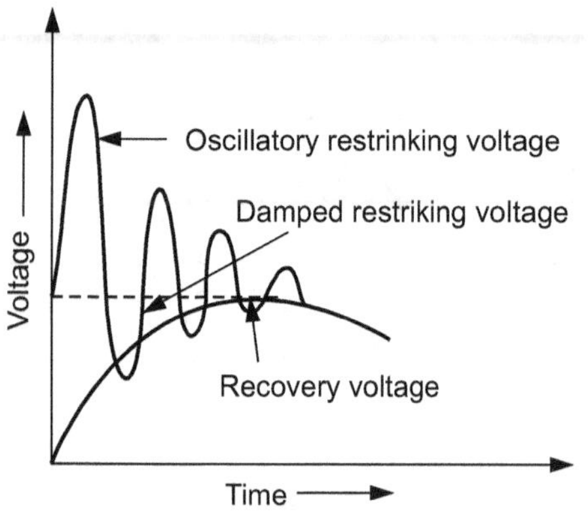

Fig. 2.8

- Referring to Fig. 2.7, when a fault occurs, the contacts of the circuit breaker are opened and an arc is struck between the contacts. Since the contacts are shunted by resistance R, a part of arc current flows through this resistance. This results in the decrease of arc current and an increase in the rate of de-ionisation of the arc path. Consequently, the arc resistance is increased.

- The increased arc resistance leads to a further increase in current through shunt resistance. This process continues until the arc current becomes so small that it fails to maintain the arc. Now, the arc is extinguished and circuit current is interrupted. The shunt resistor also helps in limiting the oscillatory growth of re-striking voltage. It can be proved mathematically that natural frequency of oscillations of the circuit shown in Fig. 2.7 is given by :

$$f_n = \frac{1}{2\pi} \sqrt{\frac{1}{LC} - \frac{1}{4R^2C^2}}$$

- The effect of shunt resistance R is to prevent the oscillatory growth of re-striking voltage and cause it to grow exponentially upto recovery voltage. This is being most effective when the value of R is so chosen that the circuit is critically damped. The value of R required for critical damping is $0.5 \sqrt{\dfrac{L}{C}}$ Fig. 2.8 shows the oscillatory growth and exponential growth when the circuit is critically damped.

- To sum up, resistors across breaker contacts may be used to perform one or more of the following functions:

  ➢ To reduce the rate of rise of re-striking voltage and the peak value of re-striking voltage.

  ➢ To reduce the voltage surges due to current chopping and capacitive current breaking.

  ➢ To ensure even sharing of re-striking voltage transient across the various breaks in multi break circuit breakers.

- It may be noted that value of resistance required to perform each function is usually different. However, it is often necessary to compromise and make one resistor do more than one of these functions.

2.6.2 Expression for RRRV

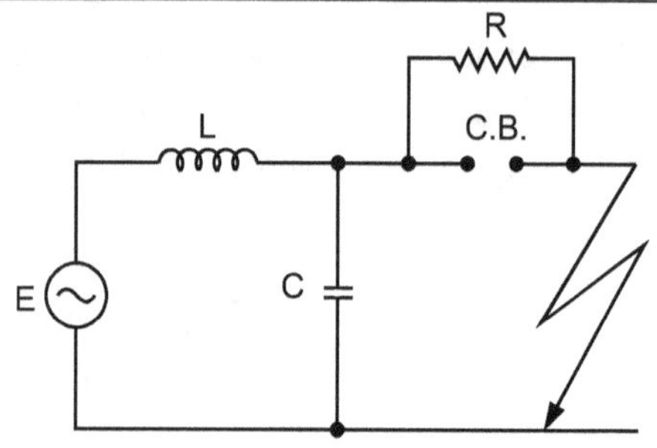

(a) Circuit

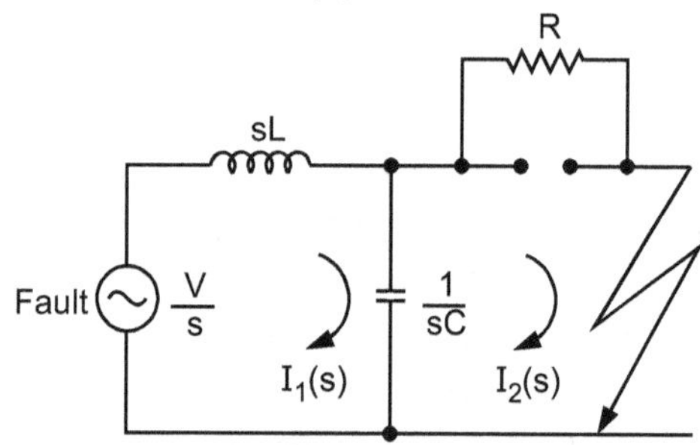

(b) Equivalent circuit

Fig. 2.9

Applying KVL to loop (I),

$$\frac{V}{s} = \left(sL + \frac{1}{sC}\right) I_1(s) - \frac{1}{sC} I_2(s) \qquad \ldots(2.1)$$

Applying KVL to loop (II),

$$0 = -\frac{1}{sC} I_1(s) + \left(R + \frac{1}{sC}\right) I_2(s) \qquad \ldots(2.2)$$

From equation (2.2),

$$I_1(s) = (1 + sCR)\, I_2(s)$$

Substituting this value in equation (2.1)

$$\frac{V}{s} = \left(sL + \frac{1}{sC}\right)(1 + sCR)\, I_2(s) - \frac{1}{sC} I_2(s)$$

$$\therefore \quad \frac{V}{s} = \left[sC + s^2RLC + \frac{1}{sC} + R - \frac{1}{sC}\right] I_2(s)$$

$$\therefore \quad \frac{V}{s} = \left[sL + s^2RLC + \frac{1}{sC} + R - \frac{1}{sC}\right] I_2(s)$$

$$\therefore \quad \frac{V}{s} = (RLCs^2 + Ls + R)\, I_2(s)$$

$$\therefore \quad I_s(s) = \frac{V}{s(RLCs^2 + Ls + R)} = \frac{V/RLC}{s\left(s^2 + \frac{1}{RC} + \frac{1}{LC}\right)}$$

Using partial fractions,

$$I_s(s) = \left\{\frac{1}{s} - \frac{s + \frac{1}{2RC}}{\left(s + \frac{1}{2RC}\right)^2 + \frac{1}{LC} - \left(\frac{1}{2RC}\right)^2} - \frac{\frac{1}{2RC}}{\left(s + \frac{1}{2RC}\right)^2 + \frac{1}{LC} - \left(\frac{1}{2RC}\right)^2}\right\}\frac{V}{R}$$

Put
$$x = \frac{1}{2RC}, \quad y = \frac{1}{LC} - \left(\frac{1}{2RC}\right)^2$$

$$I_2(s) = \frac{V}{R}\left\{\frac{1}{s}\frac{s + x}{(s + x)2 + (\sqrt{y})2} - \frac{x}{(s + x)2 + (\sqrt{y})2}\right\}$$

Taking inverse Laplace transform,

$$i_2(t) = \frac{V}{R}\left[1 - e^{-st}\left(\cos\sqrt{y}\,t + \frac{x}{y}\sin\sqrt{y}\,t\right)\right]$$

The natural frequency of oscillation is given by,

$$f_n = \frac{1}{2\pi}\sqrt{\frac{1}{LC} - \frac{1}{4C^2R^2}}$$

It can be seen that with the value of the resistance R equal to or less than $\frac{1}{2}\sqrt{\frac{L}{C}}$, the oscillatory nature of the transient will not be there and RRRV will be within the permissible limits of circuit breaker.

For critical damping

$$R = \frac{1}{2}\sqrt{\frac{L}{C}}$$

Considering different values of R, the oscillations observed are shown in the Fig. 2.10.

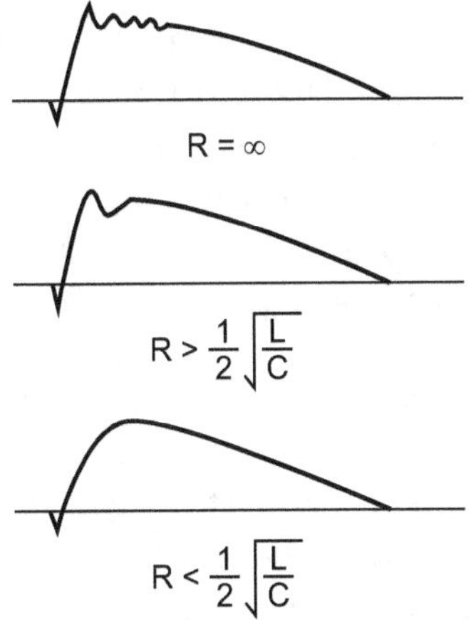

Fig. 2.10

- In air blast circuit breakers it is observed that the rate at which dielectric strength of gap increases is lower than in oil breakers. Since air has a much lower dielectric strength than the gases at same temperature and pressure in oil circuit breaker. The dielectric strength of a gas increases with pressure.

- Thus the air blast circuit breaker is more sensitive to restriking voltage transient. In low or medium voltage air blast circuit breaker the rate or rise of restriking voltage is higher. Thus shunt resistors are used for low and medium voltage air blast circuit breakers. Also in case of oil circuit breakers the resistance switching is not employed as it is not sensitive to RRRV.

SOLVED EXAMPLES

Example 2.1 : *In 132 kV transmission system, the phase to ground capacitance is 0.01 µF. The inductance being 6 H. Calculate the voltage appearing across the pole of a circuit breaker if a magnetizing current of 10 A is interrupted. Find the value of resistance to be used across contact space to eliminate the striking voltage transient.*

Solution :
$$L = 6H$$
$$C = 0.01\ \mu F = 0.01 \times 10^{-6}\ F$$
$$I = 10\ A$$

Voltage appearing across poles of circuit breaker, is given by

$$V = I\sqrt{\frac{L}{C}}$$
$$= 10\sqrt{\frac{6}{0.01 \times 10^{-6}}}$$
$$= 10\,(24494.89)$$
$$\therefore \quad V = 254\ kV$$

The value of resistance to be used across contact space is given by,

$$R = \frac{1}{2}\sqrt{\frac{L}{C}}$$
$$= \frac{1}{2}\sqrt{\frac{6}{0.01 \times 10^{-6}}}$$
$$= \frac{1}{2}\,(24494.89)$$
$$\therefore \quad R = 12.24\ k\Omega$$

2.7 CIRCUIT BREAKER RATINGS

- A circuit breaker may be called upon to operate under all conditions. However, major duties are imposed on the circuit breaker when there is a fault on the system in which it is connected. Under fault conditions, a circuit breaker is required to perform the following three duties :

1. It must be capable of opening the faulty circuit and breaking the fault current.

2. It must be capable of being closed on to a fault.

3. It must be capable of carrying fault current for a short time while another circuit breaker (in series) is clearing the fault. Corresponding to the above mentioned duties, the circuit breakers have three ratings viz.

    (i)   Breaking capacity

    (ii) Making capacity and

    (iii) Short-time capacity.

(i) Breaking Capacity :

- It is current (r.m.s.) that a circuit breaker is capable of breaking at given recovery voltage and under specified conditions (e.g., power factor, rate of rise of restriking voltage).

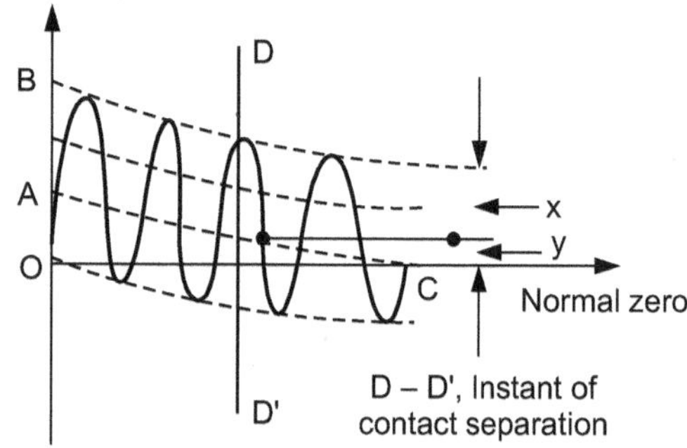

Fig. 2.11

- The breaking capacity is always stated at the r.m.s. value of fault current at the instant of contact separation. When a fault occurs, there is considerable asymmetry in the fault current due to the presence of a d.c. component. The d.c. component dies away rapidly, a typical decrement factor being 0·8 per cycle. Referring to Fig.2.11, the contacts are separated at DD′. At this instant, the fault current has

$$x = \text{maximum value of a.c. component}$$

$$y = \text{d.c. component}$$

Therefore Symmetrical breaking current = r.m.s. value of a.c. component

$$= \frac{x}{\sqrt{2}}$$

Asymmetrical breaking current = r.m.s value of total current

$$= \sqrt{\left(\frac{x^2}{\sqrt{2}} + y^2\right)}$$

- It is a common practice to express the breaking capacity in MVA by taking into account the rated breaking current and rated service voltage. Thus, if I is the rated breaking current in amperes and V is the rated service line voltage in volts, then for a 3-phase circuit, Breaking capacity $= \sqrt{3}$ V I $\times 10^{-6}$ MVA In India (or Britain), it is a usual practice to take breaking current equal to the symmetrical breaking current.

- However, American practice is to take breaking current equal to asymmetrical breaking current. Thus the American rating given to a circuit breaker is higher than the Indian or British rating. It seems to be illogical to give breaking capacity in MVA since it is obtained from the product of short-circuit current and rated service voltage.

- When the short-circuit current is flowing, there is only a small voltage across the breaker contacts, while the service voltage appears across the contacts only after the current has been interrupted. Thus MVA rating is the product of two quantities which do not exist simultaneously in the circuit.

- Therefore, the agreed international standard of specifying breaking capacity is defined as the rated symmetrical breaking current at a rated voltage.

(ii) Making Capacity :

- There is always a possibility of closing or making the circuit under shortcircuit conditions. The capacity of a breaker to "make" current depends upon its ability to withstand and close successfully against the effects of electromagnetic forces. These forces are proportional to the square of maximum instantaneous current on closing. Therefore, making capacity is stated in terms of a peak value of current instead of r.m.s. value.

- The peak value of current (including d.c. component) during the first cycle of current wave after the closure of circuit breaker is known as making capacity.

- It may be noted that the definition is concerned with the first cycle of current wave on closing the circuit breaker. This is because the maximum value of fault current possibly occurs in the first cycle only when maximum asymmetry occurs in any phase of the breaker. In other words, the making current is equal to the maximum value of asymmetrical current. To find this value, we must multiply symmetrical breaking current by $\sqrt{2}$ to convert this from r.m.s. to peak, and then by 1.8 to include the "doubling effect" of maximum asymmetry. The total multiplication factor becomes 2 . 1.8 = 2.55.

∴ Making capacity =2.55 . Symmetrical breaking capacity

(iii) Short-Time Rating :

- It is the period for which the circuit breaker is able to carry fault current while remaining closed.

- Sometimes a fault on the system is of very temporary nature and persists for 1 or 2 seconds after which the fault is automatically cleared. In the interest of continuity of supply, the breaker should not trip in such situations.

- This means that circuit breakers should be able to carry high current safely for some specified period while remaining closed i.e., they should have proven short-time rating. However, if the fault persists for a duration longer than the specified time limit, the circuit breaker will trip, disconnecting the faulty section.

- The short-time rating of a circuit breaker depends upon its ability to withstand (a) the electromagnetic force effects and (b) the temperature rise. The oil circuit breakers have a specified limit of 3 seconds when the ratio of symmetrical breaking current to the rated normal current does not exceed 40. However, if this ratio is more than 40, then the specified limit is 1 second.

- **Normal current rating.** It is the r.m.s. value of current which the circuit breaker is capable of carrying continuously at its rated frequency under specified conditions. The only limitation in this case is the temperature rise of current-carrying parts.

Example 2.2 : *A circuit breaker is rated as 1500 A, 1000 MVA, 33 kV, 3-second, 3-phase oil circuit breaker. Find (i) rated normal current (ii) breaking capacity (iii) rated symmetrical breaking current (iv) rated making current (v) short-time rating (vi) rated service voltage.*

Solution :

(i) Rated normal current = 1500 A

(ii) Breaking capacity = 1000 MVA

(iii) Rated symmetrical breaking current

$$= \frac{1000 \times 10^6}{\sqrt{3} \times 33 \times 10^3} = 17496 \text{ A (r.m.s)}$$

(iv) Rated making current = $2.55 \times 17496 = 44614$ A (peak)

(v) Short time rating = 17496 A for 3 seconds

(vi) Rated service voltage = 33 kV (r.m.s)

Example 2.3 : *A 50 Hz, 11 kV, 3-phase alternator with earthed neutral has a reactance of 5 ohms per phase and is connected to a bus-bar through a circuit breaker. The distributed capacitance upto circuit breaker between phase and neutral in 0·01 μF. Determine*

(i) *peak re-striking voltage across the contacts of the breaker*

(ii) *frequency of oscillations*

(iii) *the average rate of rise of re-striking voltage upto the first peak.*

Solution : Inductance per phase

$$L = \frac{X_L}{2\pi f} = \frac{5}{2\pi \times 50} = 0.0159 \text{ H}$$

Capacitance per phase,

$$C = 0.01 \text{ μF} = 10^{-8} \text{ F}$$

(i) Maximum value of recovery voltage (phase to neutral)

$$E_{max} = \sqrt{2} \times \frac{11}{\sqrt{3}} = 8.98 \text{ kV}$$

∴ Peak re-striking voltage = $2\, E_{max} = 2 \times 8.98 = 17.96$ kV

(ii) Frequency of oscillations is

$$f_n = \frac{1}{2\pi\sqrt{LC}} = \frac{1}{2\pi\sqrt{0.0159 \times 10^{-8}}}$$

$$= 12,628 \text{ Hz}$$

(iii) Peak re-striking voltage occurs at a time t given by

$$t = \frac{1}{f_n} = \pi\sqrt{LC} = \pi\sqrt{0.0159 \times 10^{-8}}$$

$$= 39.6 \times 10^{-6} \text{ sec} = 39.6 \text{ μ sec}$$

∴ Average rate of rise or re-string voltage

$$= \frac{\text{Peak re-strking voltage}}{\text{Time upto first peak}} = \frac{17.96 \text{ kV}}{39.6 \text{ μ sec}}$$

$$= 0.453 \text{ kV/μ sec} = 453 \times 10^3 \text{ kV/sec}$$

Example 2.4 : *In a short circuit test on a circuit breaker, the following readings were obtained on single frequency transient :*

(i) *time to reach the peak re-striking voltage, 50 μ sec*

(ii) *the peak re-striking voltage, 100 kV*

Determine the average RRRV and frequency of oscillations

Solution :

$$\text{Average RRRV} = \frac{\text{Peak re-striking voltage}}{\text{Time to reach peak value}}$$

$$= \frac{100 \text{ kV}}{50 \text{ μ sec}} = 2 \text{ k V/μ sec}$$

$$= 2 \times 10^6 \text{ kV/sec}$$

Natural frequency of oscillations,

$$f_n = \frac{1}{2 \times \text{Time to reach peak value}}$$

$$= \frac{1}{2 \times 50 \text{ μsec}}$$

$$= \frac{1}{2 \times 50 \text{ μ sec}} = 10,000 \text{ Hz}$$

Example 2.5 : *An air-blast circuit breaker is designed to interrupt a transformer magnetizing current of 11A (r.m.s.) chops the current at an instantaneous value of 7A. If the values of L and C in the circuit are 35.2 H and 0.0023 µF, find the value of voltage that appears across the contacts of the breaker. Assume that all the inductive energy is transferred to the capacitance.*

Solution : Voltage across breaker contacts at chopping is

$$e = i \sqrt{\frac{L}{C}}$$

Here , i = 7A ; L = 35.2 H and C = 0.0023 µF

$$\therefore \quad e = 7 \sqrt{\frac{35.2}{0.0023 \times 10^{-6}}} \text{ volts}$$

$$= 866 \times 10^3 \text{ V} = 866 \text{ kV}$$

2.8 AUTOMATIC RECLOSING

- About 90% of faults on overhead lines are of transient nature. Transient faults are caused by lightning or external bodies falling on the lines. Such faults are always associated with arcs. If the line is disconnected from the system for a short time, the arc is extinguished and the fault disappears. Immediately after this, the circuit breaker can be reclosed automatically to restore the supply.

- Most faults on EHV lines are caused by lightning. Flashover across insulators takes place due to overvoltages caused by lightning and exists for a short time. Hence, only on instantaneous reclosure is used in the case of EHV lines. There is no need for more than one reclosure for such a situation. For EHV lines, one reclosure in 12 cycles is recommended. A fast reclosure is desired from the stability point of view.

- On lines up 33 k V, most faults are caused by external objects such as tree branches, etc. falling on the overhead lines. This is due to the fact that the support height is less than that of the trees. The external objects may not be burnt clear at the first reclosure and may require additional reclosures. Usually three reclosures at 15-120 seconds intervals arc made to clear the fault. Statistical reports show that over 80% faults are cleared after the first reclosure, 10% require the second reclosure and 2% need the third reclosure, while the remaining 8% are permanent faults. If the fault is not cleared after 3 reclosures, it indicates that the fault is of permanent nature. Automatic reclosurc arc not used on cables as the breakdown of insulation in cables causes a permanent fault.

2.9 CLASSIFICATION OF CIRCUIT BREAKERS

There are several ways of classifying the circuit breakers. However, the most general way of classification is on the basis of medium used for arc extinction. The medium used for arc extinction is usually oil, air, sulphur hexafluoride (SF6) or vacuum. Accordingly, circuit breakers may be classified into :

- Oil circuit breakers which employ some insulating oil (e.g., Transformer oil) for arc extinction.

- Air-blast circuit breakers in which high pressure air-blast is used for extinguishing the arc.

- Sulphur hexafluroide circuit breakers in which sulphur hexafluroide (SF6) gas is used for arc extinction.

- Vacuum circuit breakers in which vacuum is used for arc extinction. Each type of circuit breaker has its own advantages and disadvantages. In the following sections, we shall discuss the construction and working of these circuit breakers with special emphasis on the way the arc extinction is facilitated.

2.10 OIL CIRCUIT BREAKERS

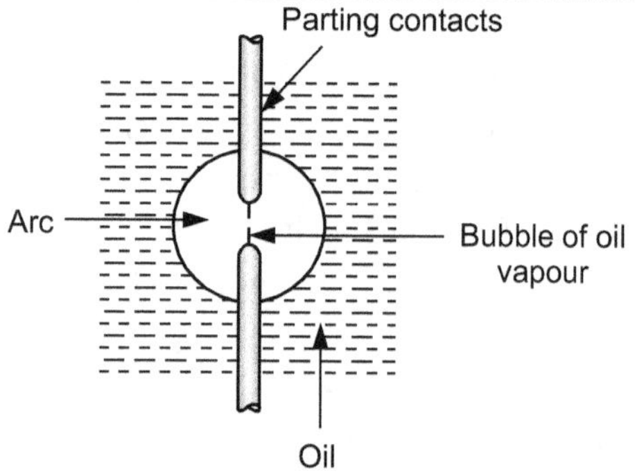

Fig. 2.12

- In such circuit breakers, some insulating oil (e.g., Transformer oil) is used as an arc quenching medium. The contacts are opened under oil and an arc is struck between them. The heat of the arc evaporates the surrounding oil and dissociates it into a substantial volume of gaseous hydrogen at high pressure.

- The hydrogen gas occupies a volume about one thousand times that of the oil decomposed. The oil is, therefore, pushed away from the arc and an expanding hydrogen gas bubble surrounds the arc region and adjacent portions of the contacts (See Fig. 2.12). The arc extinction is facilitated mainly by two processes.

- Firstly, the hydrogen gas has high heat conductivity and cools the arc, thus aiding the de-ionisation of the medium between the contacts. Secondly, the gas sets up turbulence in the oil and forces it into the space between contacts, thus eliminating the arcing products from the arc path. The result is that arc is extinguished and circuit current †interrupted.

Advantages : The advantages of oil as an arc quenching medium are:

- It absorbs the arc energy to decompose the oil into gases which have excellent cooling properties.
- It acts as an insulator and permits smaller clearance between live conductors and earthed components.
- The surrounding oil presents cooling surface in close proximity to the arc.

Disadvantages : The disadvantages of oil as an arc quenching medium are :

- It is inflammable and there is a risk of a fire.
- It may form an explosive mixture with air.
- The arcing products (e.g., carbon) remain in the oil and its quality deteriorates with successive operations. This necessitates periodic checking and replacement of oil.

2.11 TYPES OIL CIRCUIT BREAKERS

- The oil circuit breakers find extensive use in the power system. These can be classified into the following types:

1. **Bulk Oil Circuit Breakers** which use a large quantity of oil. The oil has to serve two purposes. Firstly, it extinguishes the arc during opening of contacts and secondly, it insulates the current conducting parts from one another and from the earthed tank. Such circuit breakers may be classified into :

   (a) Plain break oil circuit breakers

   (b) Arc control oil circuit breakers.

   In the former type, no special means is available for controlling the arc and the contacts are directly exposed to the whole of the oil in the tank. However, in the latter type, special arc control devices are employed to get the beneficial action of the arc as efficiently as possible.

2. **Low Oil Circuit Breakers** which use minimum amount of oil. In such circuit breakers, oil is used only for arc extinction; the current conducting parts are insulated by air or porcelain or organic insulating material.

2.11.1 Plain Break Oil Circuit Breakers

- Plain-break oil circuit breaker involves the simple process of separating the contacts under the whole of the oil in the tank. There is no special system for arc control other than the increase in length caused by the

separation of contacts. The arc extinction occurs when a certain critical gap between the contacts is reached.

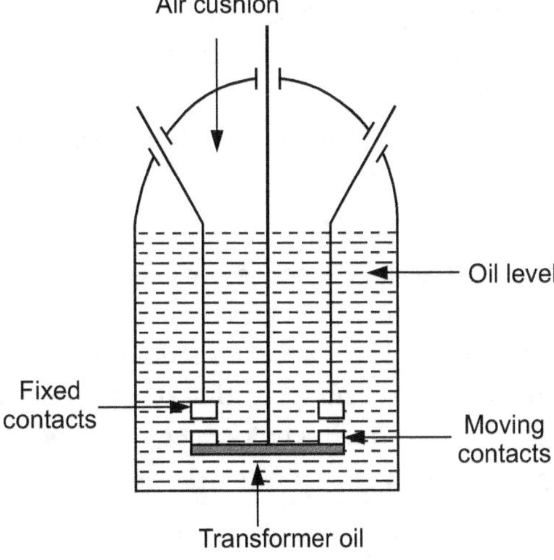

Fig. 2.13

- The plain-break oil circuit breaker is the earliest type from which all other circuit breakers have developed. It has a very simple construction. It consists of fixed and moving contacts enclosed in a strong weather-tight earthed tank containing oil upto a certain level and an air cushion above the oil level. The air cushion provides sufficient room to allow for the reception of the arc gases without the generation of unsafe pressure in the dome of the circuit breaker. It also absorbs the mechanical shock of the upward oil movement. Fig. 2.13 shows a double break plain oil circuit breaker. It is called a double break because it provides two breaks in series.

- Under normal operating conditions, the fixed and moving contacts remain closed and the breaker carries the normal circuit current. When a fault occurs, the moving contacts are pulled down by the protective system and an arc is struck which vapourises the oil mainly into hydrogen gas. The arc extinction is facilitated by the following processes :

   (i)   The hydrogen gas bubble generated around the arc cools the arc column and aids the deionization of the medium between the contacts.

   (ii) The gas sets up turbulence in the oil and helps in eliminating the arcing products from the arc path.

   (iii) As the arc lengthens due to the separating contacts, the dielectric strength of the medium is Increased. The result of these actions is that at some critical gap length, the arc is extinguished and the Circuit current is interrupted.

Disadvantages :

- There is no special control over the arc other than the increase in length by separating the Moving contacts. Therefore, for successful interruption, long arc length is necessary.

- These breakers have long and inconsistent arcing times.

- These breakers do not permit high speed interruption.

- Due to these disadvantages, plain-break oil circuit breakers are used only for low-voltage applications where high breaking-capacities are not important. It is a usual practice to use such breakers for low capacity installations for voltages not exceeding 11 kV.

2.11.2 Arc Control Oil Circuit Breakers

- In case of plain-break oil circuit breaker discussed above, there is very little artificial control over the arc. Therefore, comparatively long arc length is essential in order that turbulence in the oil caused by the gas may assist in quenching it.

- However, it is necessary and desirable that final arc extinction should occur while the contact gap is still short. For this purpose, some arc control is incorporated and the breakers are then called arc control circuit breakers. There are two types of such breakers, namely :

  1. **Self-Blast Oil Circuit Breakers :** in which arc control is provided by internal means i.e. the Arc itself is employed for its own extinction efficiently.

  2. **Forced-Blast Oil Circuit Breakers :** in which arc control is provided by mechanical means external to the circuit breaker.

1. Self-Blast Oil Circuit Breakers :

- In this type of circuit breaker, the gases produced during arcing are confined to a small volume by the use of an insulating rigid pressure chamber or pot surrounding the contacts. Since the space available for the arc gases is restricted by the chamber, a very high pressure is developed to force the oil and gas through or around the arc to extinguish it.

- The magnitude of pressure developed depends upon the value of fault current to be interrupted. As the pressure is generated by the arc itself, therefore, such breakers are sometimes called self-generated pressure oil circuit breakers.

- The pressure chamber is relatively cheap to make and gives reduced final arc extinction gap length and arcing time as against the plain-break oil circuit breaker. Several designs of pressure chambers (sometimes called explosion pots) have been developed and a few of them are described below :

(a) Plain Explosion Pot

- It is a rigid cylinder of insulating material and encloses the fixed and moving contacts (See Fig. 2.14). The moving contact is a cylindrical rod passing through a restricted opening (called throat) at the bottom. When a fault occurs, the contacts get separated and an arc is struck between them.

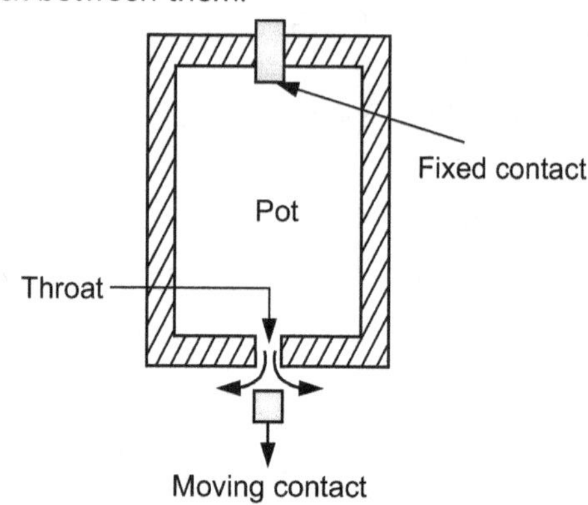

Fig. 2.14

- The heat of the arc decomposes oil into a gas at very high pressure in the pot. This high pressure forces the oil and gas through and round the arc to extinguish it. If the final arc extinction does not take place while the moving contact is still within the pot, it occurs immediately after the moving contact leaves the pot. It is because emergence of the moving contact from the pot is followed by a violent rush of gas and oil through the throat producing rapid extinction.

- The principal limitation of this type of pot is that it cannot be used for very low or for very high fault currents. With low fault currents, the pressure developed is small, thereby increasing the arcing time. On the other hand, with high fault currents, the gas is produced so rapidly that explosion pot is liable to burst due to high pressure. For this reason, plain explosion pot operates well on moderate short-circuit currents only where the rate of gas evolution is moderate.

(b) Cross Jet Explosion Pot :

- This type of pot is just a modification of plain explosion pot and is illustrated in Fig. 2.15. It is made of insulating material and has channels on one side which act as arc splitters. The arc splitters help in increasing the arc length, thus facilitating arc extinction. When a fault occurs, the moving contact of the circuit breaker begins to separate.

- As the moving contact is withdrawn, the arc is initially struck in the top of the pot. The gas generated by the arc exerts pressure on the oil in the back passage. When the moving contact uncovers the arc splitter ducts, fresh oil is forced across the arc path. The arc is, therefore, driven sideways into the "arc splitters" which increase the arc length, causing arc extinction.

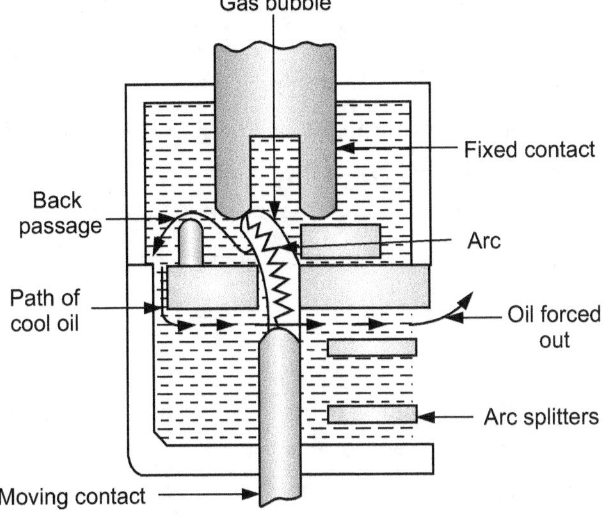

Fig. 2.15

- The cross-jet explosion pot is quite efficient for interrupting heavy fault currents. However, for low fault currents, the gas pressure is †small and consequently the pot does not give a satisfactory operation.

(c) Self-Compensated Explosion Pot :

- This type of pot is essentially a combination of plain explosion pot and cross jet explosion pot. Therefore, it can interrupt low as well as heavy short circuit currents with reasonable accuracy. Fig. 2.16 shows the schematic diagram of self-compensated explosion pot.

- It consists of two chambers, the upper chamber is the cross-jet explosion pot with two arc splitter ducts while the lower one is the plain explosion pot. When the short-circuit current is heavy, the rate of generation of gas is very high and the device behaves as a cross-jet explosion pot.

- The arc extinction takes place when the moving contact uncovers the first or second arc splitter duct. However, on low short-circuit currents, the rate of gas generation is small and the tip of the moving contact has the time to reach the lower chamber. During this

time, the gas builds up sufficient pressure as there is very little leakage through arc splitter ducts due to the obstruction offered by the arc path and right angle bends. When the moving contact comes out of the throat, the arc is extinguished by plain pot action. It may be noted that as the severity of the short-circuit current increases, the device operates less and less as a plain explosion pot and more and more as a cross-jet explosion pot. Thus the tendency is to make the control self-compensating over the full range of fault currents to be interrupted.

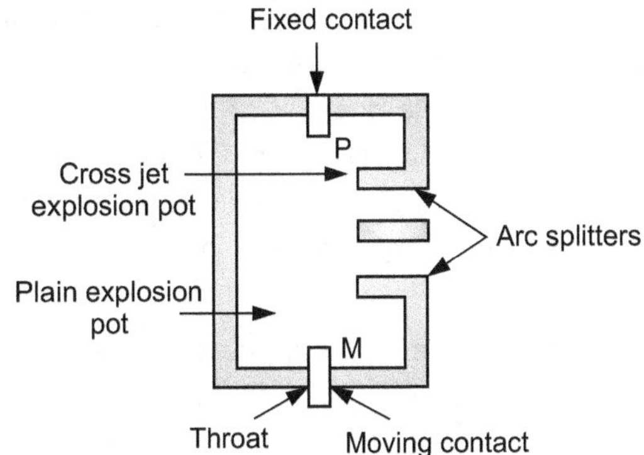

Fig. 2.16

2. Forced-Blast Oil Circuit Breakers :

- In the self-blast oil circuit breakers discussed above, the arc itself generates the necessary pressure to force the oil across the arc path. The major limitation of such breakers is that arcing times tend to be long and inconsistent when operating against currents considerably less than the rated currents. It is because the gas generated is much reduced at low values of fault currents.

- This difficulty is overcome in forced-blast oil circuit breakers in which the necessary pressure is generated by external mechanical means independent of the fault currents to be broken. In a forced -blast oil circuit breaker, oil pressure is created by the piston-cylinder arrangement.

- The movement of the piston is mechanically coupled to the moving contact. When a fault occurs, the contacts get separated by the protective system and an arc is struck between the contacts. The piston forces a jet of oil towards the contact gap to extinguish the arc. It may be noted that necessary oil pressure produced does not in any way depend upon the fault current to be broken.

Advantages :

- Since oil pressure developed is independent of the fault current to be interrupted, the performance at low currents is more consistent than with self-blast oil circuit breakers.

- The quantity of oil required is reduced considerably.

2.11.3 Low Oil Circuit Breakers

- In the bulk oil circuit breakers discussed so far, the oil has to perform two functions. Firstly, it acts as an arc quenching medium and secondly, it insulates the live parts from earth. It has been found that only a small percentage of oil is actually used for arc extinction while the major part is utilized for insulation purposes. For this reason, the quantity of oil in bulk oil circuit breakers reaches a very high figure as the system voltage increases. This not only increases the expenses, tank size and weight of the breaker but it also increases the fire risk and maintenance problems.

- The fact that only a small percentage of oil (about 10% of total) in the bulk oil circuit breaker is actually used for arc extinction leads to the question as to why the remainder of the oil, that is not immediately surrounding the device, should not be omitted with consequent saving in bulk, weight and fire risk.

- This led to the development of low-oil circuit breaker. A low oil circuit breaker employs solid materials for insulation purposes and uses a small quantity of oil which is just sufficient for arc extinction. As regards quenching the arc, the oil behaves identically in bulk as well as low oil circuit breaker. By using suitable arc control devices, the arc extinction can be further facilitated in a low oil circuit breaker.

Construction :

- Fig 2.17 shows the cross section of a single phase low oil circuit breaker. There are two compartments separated from each other but both filled with oil. The upper chamber is the circuit breaking chamber while the lower one is the supporting chamber. The two chambers are separated by a partition and oil from one chamber is prevented from mixing with the other chamber.

- This arrangement permits two advantages. Firstly, the circuit breaking chamber requires a small volume of oil which is just enough for arc extinction. Secondly, the amount of oil to be replaced is reduced as the oil in the supporting chamber does not get contaminated by the arc.

1. **Supporting Chamber :** It is a porcelain chamber mounted on a metal chamber. It is filled with oil which is physically separated from the oil in the circuit breaking compartment. The oil inside the supporting chamber and the annular space formed between the porcelain insulation and bakelised paper is employed for insulation purposes only.

2. **Circuit-Breaking Chamber :** It is a porcelain enclosure mounted on the top of the supporting compartment. It is filled with oil and has the following parts :

   (a) Upper and lower fixed contacts

   (b) Moving contact

   (c) Turbulator

   The moving contact is hollow and includes a cylinder which moves down over a fixed piston. The turbulator is an arc control device and has both axial and radial vents. The axial venting ensures the interruption of low currents whereas radial venting helps in the interruption of heavy currents.

3. **Top Chamber :** It is a metal chamber and is mouted on the circuit-breaking chamber. It provides expansion space for the oil in the circuit breaking compartment. The top chamber is also provided with a separator which prevents any loss of oil by centrifugal action caused by circuit breaker operation during fault conditions.

Operation : Under normal operating conditions, the moving contact remains engaged with the upper fixed contact. When a fault occurs, the moving contact is pulled down by the tripping springs and an arc is struck. The arc energy vaporises the oil and produces gases under high pressure. This action constrains the oil to pass through a central hole in the moving contact and results in forcing series of oil through the respective passages of the turbulator. The process of turbulation is orderly one, in which the sections of the arc are successively quenched by the effect of separate streams of oil moving across each section in turn and bearing away its gases.

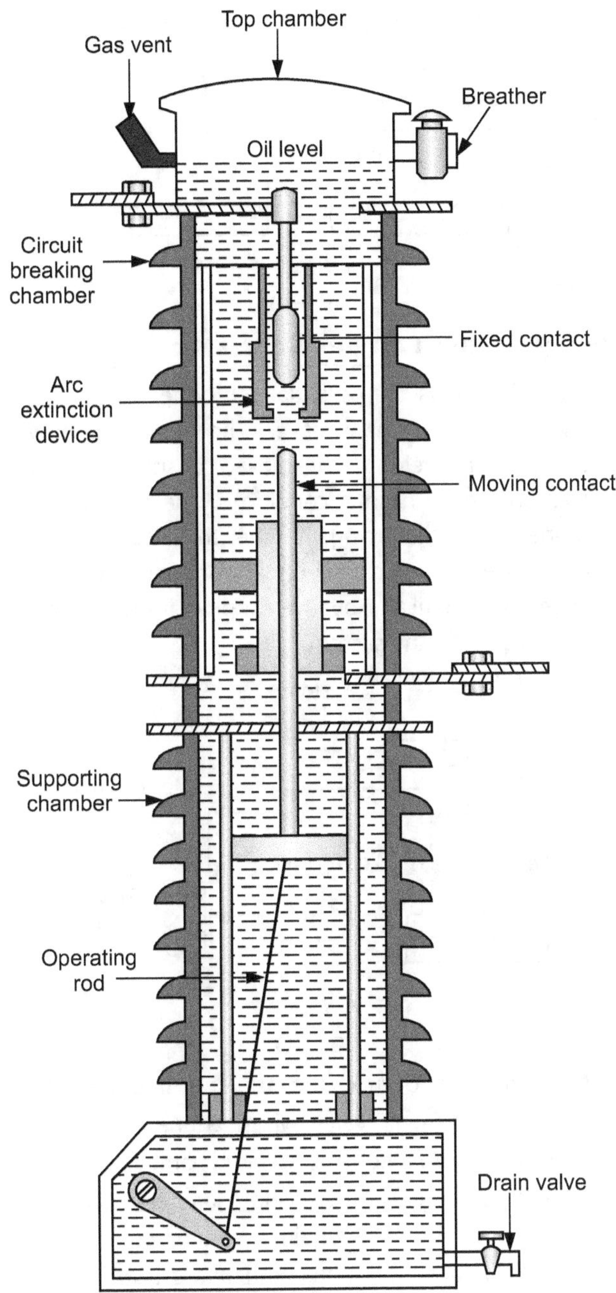

Fig. 2.17

Advantages : A low oil circuit breaker has the following advantages over a bulk oil circuit breaker:

- It requires lesser quantity of oil.
- It requires smaller space.
- There is reduced risk of fire.
- Maintenance problems are reduced.

Disadvantages : A low oil circuit breaker has the following disadvantages as compared to a bulk oil circuit breaker :

- Due to smaller quantity of oil, the degree of carbonisation is increased.
- There is a difficulty of removing the gases from the contact space in time.

- The dielectric strength of the oil deteriorates rapidly due to high degree of carbonization.

2.12 MAINTENANCE OF OIL CIRCUIT BREAKERS

- The maintenance of oil circuit breaker is generally concerned with the checking of contacts and Dielectric strength of oil. After a circuit breaker has interrupted fault currents a few times or load currents several times, its contacts may get burnt by arcing and the oil may lose some of its dielectric strength due to carbonisation.
- This results in the reduced rupturing capacity of the breaker. Therefore, it is a good practice to inspect the circuit breaker at regular intervals of 3 or 6 months. During inspection of the breaker, the following points should be kept in view:
  - ➢ Check the current carrying parts and arcing contacts. If the burning is severe, the contacts should be replaced.
  - ➢ Check the dielectric strength of the oil. If the oil is badly discoloured, it should be changed or reconditioned. The oil in good condition should withstand 30 kV for one minute in a standard oil testing cup with 4 mm gap between electrodes.
  - ➢ Check the insulation for possible damage. Clean the surface and remove carbon deposits with a strong and dry fabric.
  - ➢ Check the oil level.
  - ➢ Check closing and tripping mechanism

2.13 AIR BLAST CIRCUIT BREAKERS

- These breakers employ a high pressure air-blast as an arc quenching medium. The contacts are opened in a flow of air-blast established by the opening of blast valve. The air-blast cools the arc and sweeps away the arcing products to the atomsphere.
- This rapidly increases the dielectric strength of the medium between contacts and prevents from re-establishing the arc. Consequently, the arc is extinguished and flow of current is interrupted.

Advantages : An air-blast circuit breaker has the following advantages over an oil circuit breaker

- The risk of fire is eliminated.
- The arcing products are completely removed by the blast whereas the oil deteriorates with successive operations; the expense of regular oil replacement is avoided.
- The growth of dielectric strength is so rapid that final contact gap needed for arc extinction is very small. This reduces the size of the device.

- The arcing time is very small due to the rapid build up of dielectric strength between contacts. Therefore, the arc energy is only a fraction of that in oil circuit breakers, thus resulting in less burning of contacts.
- Due to lesser arc energy, air-blast circuit breakers are very suitable for conditions where frequent operation is required.
- The energy supplied for arc extinction is obtained from high pressure air and is independent of the current to be interrupted.

Disadvantages : The use of air as the arc quenching medium offers the following disadvantges :

- The air has relatively inferior arc extinguishing properties.
- The air-blast circuit breakers are very sensitive to the variations in the rate of rise of restriking voltage.
- Considerable maintenance is required for the compressor plant which supplies the air-blast.
- The air blast circuit breakers are finding wide applications in high voltage installations. Majority of the circuit breakers for voltages beyond 110 kV are of this type.

2.14 TYPES OF AIR-BLAST CIRCUIT BREAKERS

Depending upon the direction of air-blast in relation to the arc, air-blast circuit breakers are classified in to:

1. Axial-Blast Type

- Axial-blast type in which the air-blast is directed along the arc path as shown in Fig. 2.18 below

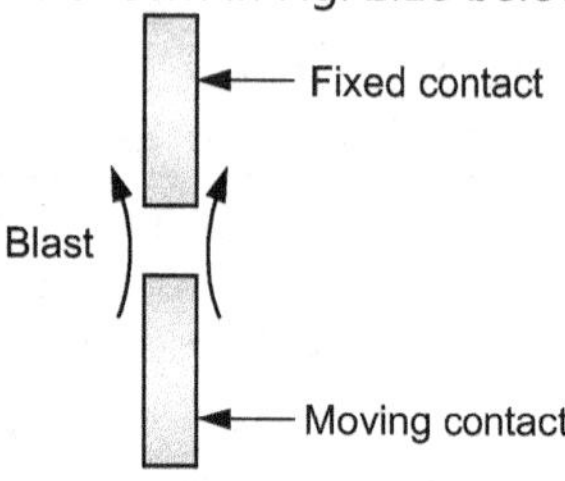

Fig. 2.18

2. Cross-Blast Type

- Cross-blast type in which the air-blast is directed at right angles to the arc path as shown in Fig. 2.19 below

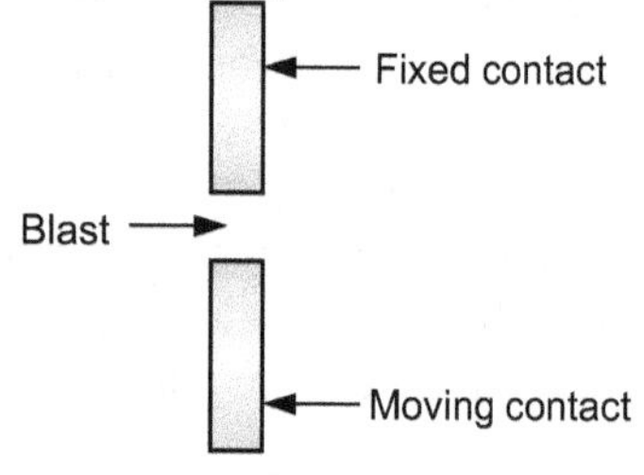

Fig. 2.19

3. Radial-blast type

- Radial-blast type in which the air-blast is directed radially as shown in Fig. 2.20 below.

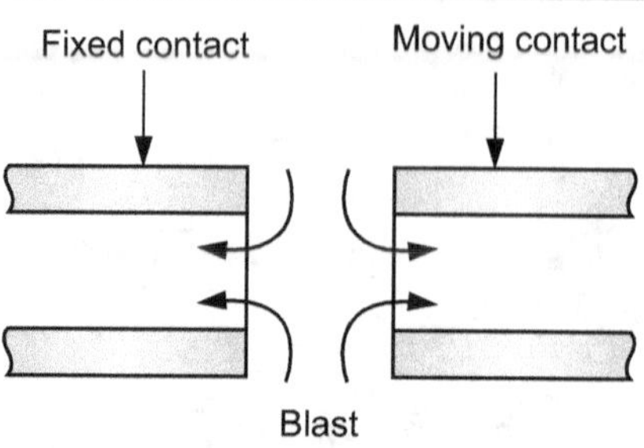

Fig. 2.20

2.14.1 Axial-Blast Air Circuit Breaker

- Fig 2.21 shows the essential components of a typical axial-blast air circuit breaker. The fixed and moving contacts are held in the closed position by spring pressure under normal conditions. The air reservoir is connected to the arcing chamber through an air valve. This valve remains closed under normal conditions but opens automatically by the tripping impulse when a fault occurs on the system.

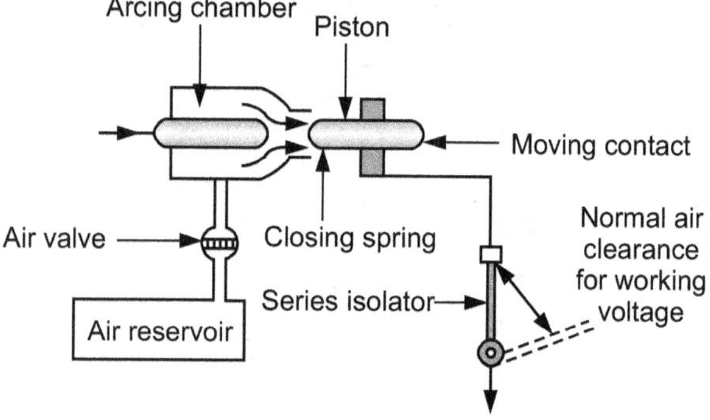

Fig. 2.21

- When a fault occurs, the tripping impulse causes opening of the air valve which connects the circuit breaker reservoir to the arcing chamber. The high pressure air entering the arcing chamber pushes away the moving contact against spring pressure. The moving contact is separated and an arc is struck. At the same time, high pressure air blast flows along the arc and takes away the ionized gases along with it. Consequently, the arc is extinguished and current flow is interrupted.

- It may be noted that in such circuit breakers, the contact separation required for interruption is generally small (1.75 cm or so). Such a small gap may constitute inadequate clearance for the normal service voltage. Therefore, an isolating switch is incorporated as a part of this type of circuit breaker. This switch opens immediately after fault interruption to provide the necessary clearance for insulation.

2.14.2 Cross-Blast Air Breaker

- In this type of circuit breaker, an air-blast is directed at right angles to the arc. The cross-blast lengthens and forces the arc into a suitable chute for arc extinction. Fig. 2.22 shows the essential parts of a typical cross-blast air circuit breaker. When the moving contact is withdrawn, an arc is struck between the fixed and moving contacts. The high pressure cross-blast forces the arc into a chute consisting of arc splitters and baffles.

- The splitters serve to increase the length of the arc and baffles give improved cooling. The result is that arc is extinguished and flow of current is interrupted. Since blast pressure is same for all currents, the inefficiency at low currents is eliminated. The final gap for interruption is great enough to give normal insulation clearance so that a series isolating switch is not necessary.

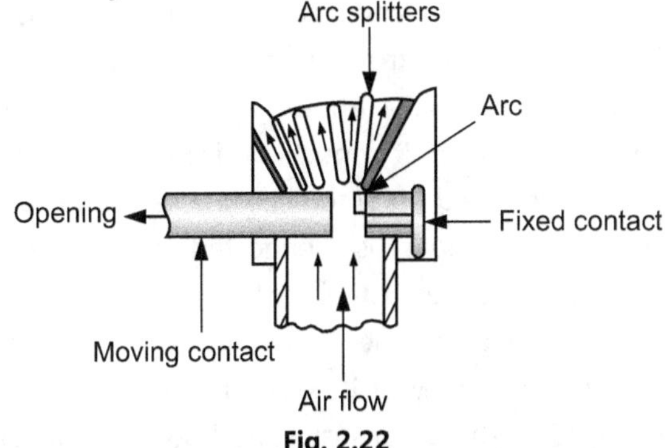

Fig. 2.22

2.15 SULPHUR HEXAFLOURIDE (SF6) CIRCUIT BREAKERS

- In such circuit breakers, sulphur hexaflouride (SF6) gas is used as the arc quenching medium. The SF6 is an electro-negative gas and has a strong tendency to absorb free electrons. The contacts of the breaker are opened in a high pressure flow of SF6 gas and an arc is struck between them.

- The conducting free electrons in the arc are rapidly captured by the gas to form relatively immobile negative ions. This loss of conducting electrons in the arc quickly builds up enough insulation strength to extinguish the arc. The SF6 circuit breakers have been found to be very effective for high power and high voltage service.

Construction :

- Fig. 2.23 shows the parts of a typical SF6 circuit breaker. It consists of fixed and moving contacts enclosed in a chamber (called arc interruption chamber) containing SF6 gas. This chamber is connected to SF6 gas reservoir. When the contacts of breaker are opened, the valve mechanism permits a high pressure SF6 gas from the reservoir to flow towards the arc interruption chamber.

- The fixed contact is a hollow cylindrical current carrying contact fitted with an arc horn. The moving contact is also a hollow cylinder with rectangular holes in the sides to permit the SF6 gas to let out through these holes after flowing along and across the arc. The tips of fixed contact, moving contact and arcing horn are coated with copper-tungsten arc resistant material. Since SF6 gas is costly, it is reconditioned and reclaimed by suitable auxiliary system after each operation of the breaker.

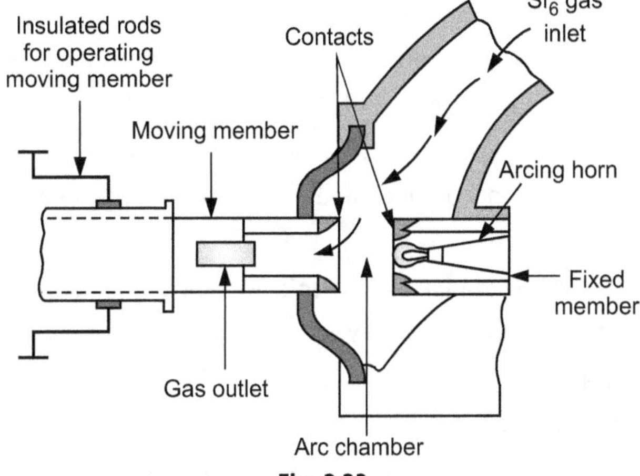

Fig. 2.23

Working: In the closed position of the breaker, the contacts remain surrounded by SF6 gas at a pressure of about 2.8 kg/cm^2. When the breaker operates, the moving contact is pulled apart and an arc is struck between the contacts. The movement of the moving contact is synchronised with the opening of a valve which permits SF6 gas at 14 kg/cm^2 pressure from the reservoir to the arc interruption chamber. The high pressure flow of SF6 rapidly absorbs the free electrons in the arc path to form immobile negative ions which are ineffective as charge carriers. The result is that the medium between the contacts quickly builds up high dielectric strength and causes the extinction of the arc. After the breaker operation (i.e., after arc extinction), the valve is closed by the action of a set of springs.

Advantages : Due to the superior arc quenching properties of SF6 gas, the SF6 circuit breakers have many advantages over oil or air circuit breakers. Some of them are listed below :

- Due to the superior arc quenching property of SF6, such circuit breakers have very short arcing time.

- Since the dielectric strength of SF6 gas is 2 to 3 times that of air, such breakers can interrupt much larger currents.

- The SF6 circuit breaker gives noiselss operation due to its closed gas circuit and no exhaust to atmosphere unlike the air blast circuit breaker.

- The closed gas enclosure keeps the interior dry so that there is no moisture problem.

- There is no risk of fire in such breakers because SF6 gas is non-inflammable.

- There are no carbon deposits so that tracking and insulation problems are eliminated.

- The SF6 breakers have low maintenance cost, light foundation requirements and minimum auxiliary equipment.

- Since SF6 breakers are totally enclosed and sealed from atmosphere, they are particularly suitable where explosion hazard exists e.g., coal mines.

Disadvantages :

- SF6 breakers are costly due to the high cost of SF6.

- Since SF6 gas has to be reconditioned after every operation of the breaker, additional equipment is requried for this purpose.

Applications :

- A typical SF6 circuit breaker consists of interrupter units each capable of dealing with currents upto 60 kA and voltages in the range of 50-80 kV. A number of units are connected in series according to the system voltage. SF6 circuit breakers have been developed for voltages 115 kV to 230 kV, power ratings 10 MVA to 20 MVA and interrupting time less than 3 cycles.

2.16 VACUUM CIRCUIT BREAKERS(VCB)

- In such breakers, vacuum (degree of vacuum being in the range from ($10-7$ to $10-5$ torr) is used as the arc quenching medium. Since vacuum offers the highest insulating strength, it has far superior arc quenching properties than any other medium.

- For example, when contacts of a breaker are opened in vacuum, the interruption occurs at first current zero with dielectric strength between the contacts building up at a rate thousands of times higher than that obtained with other circuit breakers.

Principle :

- The production of arc in a vacuum circuit breaker and its extinction can be explained as follows : When the contacts of the breaker are opened in vacuum ($10-7$ to $10-5$ torr), an arc is produced between the contacts by the ionisation of metal vapours of contacts.

- However, the arc is quickly extinguished because the metallic vapours, electrons and ions produced during arc rapidly condense on the surfaces of the circuit breaker contacts, resulting in quick recovery of dielectric strength. As soon as the arc is produced in vacuum, it is quickly extinguished due to the fast rate of recovery of dielectric strength in vacuum.

Construction : Fig. 2.24 shows the parts of a typical vacuum circuit breaker. It consists of fixed contact, moving contact and arc shield mounted inside a vacuum chamber. The movable member is connected to the control mechanism by stainless steel bellows. This enables the permanent sealing of the vacuum chamber so as to eliminate the possibility of leak. A glass vessel or ceramic vessel is used as the outer insulating body. The arc shield prevents the deterioration of the internal dielectric strength by preventing metallic vapours falling on the inside surface of the outer insulating cover.

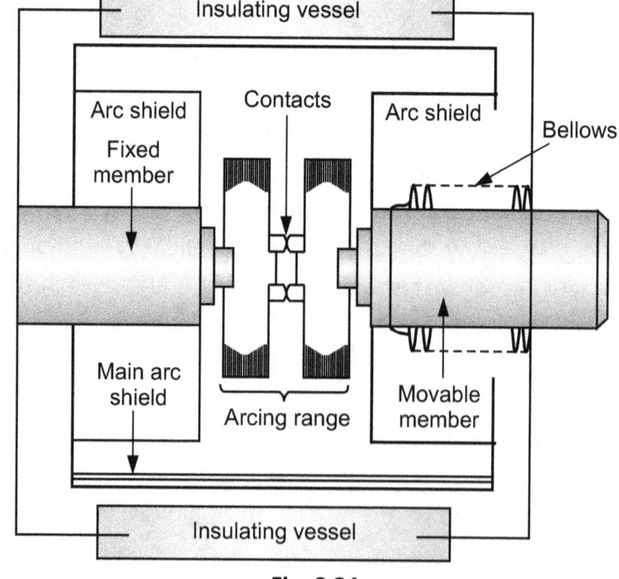

Fig. 2.24

Working :

- When the breaker operates, the moving contact separates from the fixed contact and an arc is struck between the contacts. The production of arc is due to the ionisation of metal ions and depends very much upon the material of contacts.
- The arc is quickly extinguished because the metallic vapours, electrons and ions produced during arc are diffused in a short time and seized by the surfaces of moving and fixed members and shields. Since vacuum has very fast rate of recovery of dielectric strength, the arc extinction in a vacuum breaker occurs with a short contact separation (say 0·625 cm).

Advantages : Vacuum circuit breakers have the following advantages :

- They are compact, reliable and have longer life.
- There are no fire hazards.
- There is no generation of gas during and after operation.
- They can interrupt any fault current. The outstanding feature of a VCB is that it can break any heavy fault current perfectly just before the contacts reach the definite open position.
- They require little maintenance and are quiet in operation.
- They can successfully withstand lightning surges.
- They have low arc energy.
- They have low inertia and hence require smaller power for control mechanism.

Applications :

- For a country like India, where distances are quite large and accessibility to remote areas difficult, the installation of such outdoor, maintenance free circuit breakers should prove a definite advantage.
- Vacuum circuit breakers are being employed for outdoor applications ranging from 22 kV to 66 kV. Even with limited rating of say 60 to 100 MVA, they are suitable for a majority of applications in rural areas.

2.17   TYPES OF FUSES

- Fuse is the simplest current interrupting device for protection against excessive currents. Since the invention of first fuse by Edison, several improvements have been made and now-a-days, a variety of fuses are available. Some fuses also incorporate means for extinguishing the arc that appears when the fuse element melts. In general, fuses may be classified into :
  1. Low voltages fuses
  2. High voltage fuses

- It is a usual practice to provide isolating switches in series with fuses where it is necessary to permit fuses to be replaced or rewired with safety. If such means of isolation are not available, the fuses must be so shielded as to protect the user against accidental contact with the live metal when the fuse carrier is being inserted or removed.

2.18 LOW VOLTAGE FUSES

Low voltage fuses can be subdivided into two classes viz.,

1. Semi-Enclosed Rewireable Fuse
2. High Rupturing Capacity (H.R.C.) cartridge fuse.

1.   Semi-Enclosed Rewireable Fuse :

- Rewireable fuse (also known as kit-kat type) is used where low values of fault current are to be interrupted. It consists of (i) a base and (ii) a fuse carrier. The base is of porcelain and carries the fixed contacts to which the incoming and outgoing phase wires are connected. The fuse carrier is also of porcelain and holds the fuse element (tinned copper wire) between its terminals. The fuse carrier can be inserted in or taken out of the base when desired.

- When a fault occurs, the fuse element is blown out and the circuit is interrupted. The fuse carrier is taken out and the blown out fuse element is replaced by the new one. The fuse carrier is then reinserted in the base to restore the supply. This type of fuse has two advantages. Firstly, the detachable fuse carrier permits the replacement of fuse element without any danger of coming in contact with live parts. Secondly, the cost of replacement is negligible.

Disadvantages :

- There is a possibility of renewal by the fuse wire of wrong size or by improper material.
- This type of fuse has a low-breaking capacity and hence cannot be used in circuits of high fault level.
- The fuse element is subjected to deterioration due to oxidation through the continuous heating up of the element. Therefore, after some time, the current rating of the fuse is decreased
  i.e., the fuse operates at a lower current than originally rated.
- The protective capacity of such a fuse is uncertain as it is affected by the ambient conditions.
- Accurate calibration of the fuse wire is not possible because fusing current very much depends upon the length of the fuse element. Semi-enclosed rewireable

fuses are made upto 500 A rated current, but their breaking capacity islow e.g., on 400 V service, the breaking capacity is about 4000 A. Therefore, the use of this type of fuses is limited to domestic and lighting loads.

2. **High-Rupturing Capacity (H.R.C.) Cartridge Fuse :**
   The primary objection of low and uncertain breaking capacity of semi-enclosed rewireable fuses is overcome in H.R.C. cartridge fuse. Fig. 2.25 shows the essential parts of a typical H.R.C. cartridge fuse. It consists of a heat resisting ceramic body having metal end-caps to which is welded silver current-carrying element. The space within the body surrounding the element is completely packed with a filling powder. The filling material may be chalk, plaster of paris, quartz or marble dust and acts as an arc quenching and cooling medium.

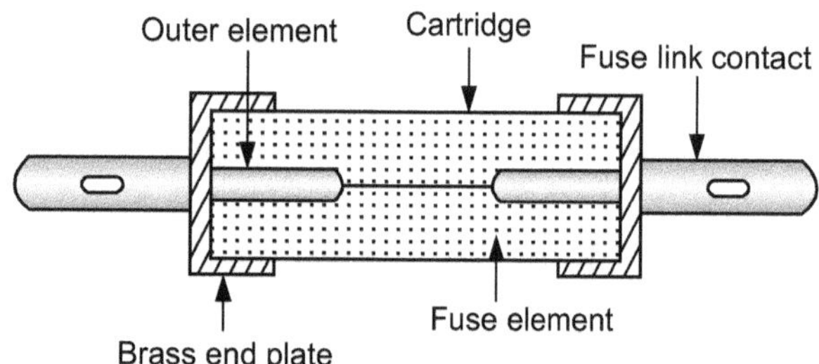

Fig. 2.25

- Under normal load conditions, the fuse element is at a temperature below its melting point. Therefore, it carries the normal current without overheating. When a fault occurs, the current increases and the fuse element melts before the fault current reaches its first peak. The heat produced in the process vapourises the melted silver element. The chemical reaction between the silver vapour and the filling powder results in the formation of a high resistance substance which helps in quenching the arc.

Advantages

- They are capable of clearing high as well as low fault currents.

- They do not deteriorate with age.

- They have high speed of operation.

- They provide reliable discrimination.

- They require no maintenance.

- They are cheaper than other circuit interrupting devices of equal breaking capacity.

- They permit consistent performance.

Disadvantages

- They have to be replaced after each operation.

- Heat produced by the arc may affect the associated switches.

3. **H.R.C. Fuse with Tripping Device :**

- Sometime, H.R.C. cartridge fuse is provided with a tripping device. When the fuse blows out under fault conditions, the tripping device causes the circuit breaker to operate. Fig. 2.26 shows the essential parts of a H.R.C. fuse with a tripping device. The body of the fuse is of ceramic material with a metallic cap rigidly fixed at each end.

- These are connected by a number of silver fuse elements. At one end is a plunger which under fault conditions hits the tripping mechanism of the circuit breaker and causes it to operate. The plunger is electrically connected through a fusible link, chemical charge and a tungsten wire to the other end of the cap as shown.

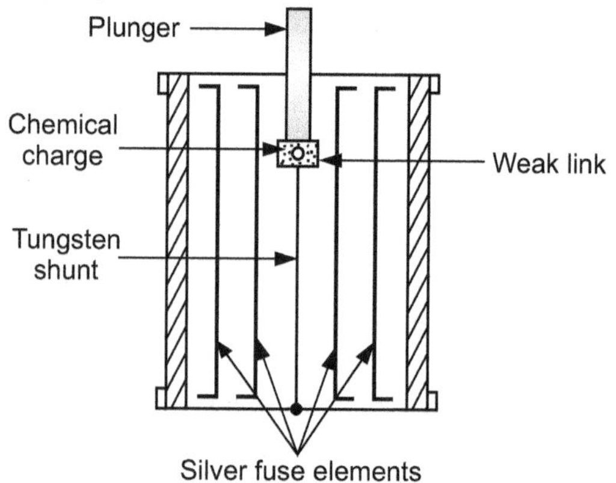

Fig. 2.26

- When a fault occurs, the silver fuse elements are the first to be blown out and then current is transferred to the tungsten wire. The weak link in series with the tungsten wire gets fused and causes the chemical charge to be detonated. This forces the plunger outward to operate the circuit breaker. The travel of the plunger is so set that it is not ejected from the fuse body under fault conditions.

Advantages : H.R.C. fuse with a tripping device has the following advantages over a H.R.C. fuse without tripping device :

- In case of a single phase fault on a three-phase system, the plunger operates the tripping mechanism of circuit breaker to open all the three phases and thus prevents "single phasing".

- The effects of full short circuit current need not be considered in the choice of circuit breaker. This permits the use of a relatively inexpensive circuit breaker.

- The fuse-tripped circuit breaker is generally capable of dealing with fairly small fault currents itself. This avoids the necessity for replacing the fuse except after highest currents for which it is intended.

Low voltage H.R.C. fuses may be built with a breaking capacity of 16,000 A to 30,000 A at 440V. They are extensively used on low-voltage distribution system against over-load and shortcircuit conditions.

2.19 HIGH VOLTAGE FUSES

- The low-voltage fuses discussed so far have low normal current rating and breaking capacity. Therefore, they cannot be successfully used on modern high voltage circuits. Intensive research by the manufacturers and supply engineers has led to the development of high voltage fuses.

Some of the high voltage fuses are :

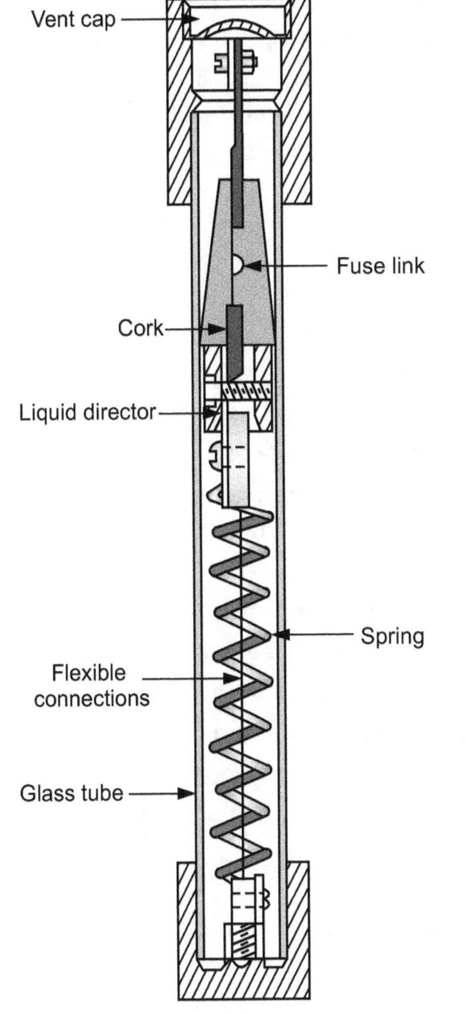

Fig. 2.27

(i) Cartridge Type :

- This is similar in general construction to the low voltage cartridge type except that special design features are incorporated. Some designs employ fuse elements wound in the form of a helix so as to avoid corona effects at higher voltages. On some designs, there are two fuse elements in parallel ; one of low resistance (silver wire) and the other of high resistance (tungsten wire).

- Under normal load conditions, the low resistance element carries the normal current. When a fault occurs, the low-resistance element is blown out and the high resistance element reduces the short-circuit current and finally breaks the circuit. High voltage cartridge fuses are used upto 33 kV with breaking capacity of about 8700 A at that voltage. Rating of the order of 200 A at 6·6 kV and 11 kV and 50 A at 33 kV are also available.

(ii) Liquid Type :

- These fuses are filled with carbon tetrachloride and have the widest range of application to h.v. systems. They may be used for circuits upto about 100 A rated current on systems upto 132 Kv and may have breaking capacities of the order of 6100 A.

- Fig. 2.27 shows the essential parts of the liquid fuse. It consists of a glass tube filled with carbon tetrachloride solution and sealed at both ends with brass caps.

- The fuse wire is sealed at one end of the tube and the other end of the wire is held by a strong phosphor bronze spiral spring fixed at the other end of the glass tube. When the current exceeds the prescribed limit, the fuse wire is blown out.

- As the fuse melts, the spring retracts part of it through a baffle (or liquid director) and draws it well into the liquid. The small quantity of gas generated at the point of fusion forces some part of liquid into the passage through baffle and there it effectively extinguishes the arc.

(iii) Metal Clad Fuses :

- Metal clad oil-immersed fuses have been developed with the object of providing a substitute for the oil circuit breaker. Such fuses can be used for very high voltage circuits and operate most satisfactorily under short-circuit conditions approaching their rated capacity.

2.20 CURRENT CARRYING CAPACITY OF FUSE ELEMENT

The current carrying capacity of a fuse element mainly depends on the metal used and the cross-sectional area but is affected also by the length, the state of surface and the surroundings of the fuse. When the fuse element attains steady temperature,

Heat produced per sec = Heat lost per second by convection, ration and conduction

or $\qquad I^2R$ = Constant × Effective surface area

or $\qquad I^2 \left(\rho \dfrac{l}{a} \right)$ = constant × d × l

where $\qquad d$ = diameter of fuse element

$\qquad\qquad l$ = length of fuse element

$\therefore \qquad I^2 \dfrac{\rho l}{(\pi/4)\, d^2}$ = constant × d × l

$\qquad$ or $\qquad l^2$ = constant × d^3

$\qquad$ or $\qquad l^2 \propto d^3$ $\qquad\qquad$ ...(2.3)

Expression (i) is known as ordinary fuse law.

2.20.1 Difference Between a Fuse and Circuit Breaker

It is worthwhile to indicate the salient differences between a fuse and a circuit breaker in the tabular form.

Sr. No.	Particular	Fuse	Circular Breaker
1.	Function	It performs both detection and interruption functions.	It performs interruption function only. The detection of fault is made by relay system.
2.	Operation	Inherently completely automatic.	Requires elaborate equipment (i.e. relays) for automatic action.
3.	Breaking capacity	Small	Very large.
4.	Operating time	Very small (0.002 sec or so)	Comparatively large (0.1 to 0.2 sec)
5.	Replacement	Requires replacement after every operation.	No replacement after operation.

1. What is a circuit breaker? Describe its operating principle.

2. Discuss the arc phenomenon in a circuit breaker.

3. Explain the various methods of arc extinction in a circuit breaker.

4. Define and explain the following terms as applied to circuit breakers.
   (i) Arc voltage
   (ii) Restriking voltage
   (iii) Recovery voltage

5. Describe briefly the action of an oil circuit breaker. How does oil help in arc extinction?

6. What are the important components common to most of circuit breakers? Discuss each component briefly.

7. Write a short note on the rate of re-striking voltage indicating its importance in the arc extinction.

8. Explain the difference between bulk oil circuit breakers and low-oil circuit breakers.

9. Discuss the constructional details and operation of a typical low-oil circuit breaker What are its relative merits and demerits?

10. Discuss the principle of operation of an air-blast circuit breaker. What are the advantages and disadvantage of using air as the arc quenching medium ?

11. Explain briefly the following types of air-blast circuit breakers :
    (i) Axial-blast type
    (ii) Cross-blast type

12. Write a short note on the rate of re-striking voltage indicating its importance in the arc extinction.

13. Discuss the phenomenon of
    (i) Current chopping
    (ii) Capacitive current breaking

14. Write short notes on the following :
    (i) Resistance switching
    (ii) Circuit breaker ratings
    (iii) Circuit interruption problems

◇ ◇ ◇

BASIC ELEMENTS OF DIGITAL PROTECTION

3.1 INTRODUCTION

- Operating voltages and currents flowing through a power system are usually at kilovolt and kiloampere levels. However, for digital processing, it is necessary to reduce the primary measurands to manageable levels. Therefore, the analogue signals are converted to digital form, thereby allowing subsequent digital processing to be performed to determine the circuit state.

- In this Chapter the basic principles underlying the conversion of analogue signals into equivalent digital forms will be explained. We shall also explain the essentially common features of various digital relaying schemes, other detailed aspects being discussed in later chapters.

3.2 BASIC COMPONENTS OF A DIGITAL RELAY

- Any digital relay can be thought of as comprising three fundamental subsystems (Figure 3.1):

    (i)   a signal conditioning subsystem

    (ii) a conversion subsystem

    (iii) a digital processing relay subsystem.

- The first two subsystems are generally common to all digital protective schemes, while the third varies according to the application of a particular scheme. Each of the three subsystems is built up of a number of components and circuits, as discussed in detail in the following sections.

3.2.1 Different Components of a Digital Relay

- Isolation transformer and surge protection circuit
- Multiplexor and S/H circuits
- Anti Aliasing Filters
- Digital Input and Output systems.
- Central Processing unit
- Event Storage system
- Signal conditioning circuit.
- Communication Peripherals
- Power Supply Block
- Sampling Clock

Isolation Transformer and Surge Protection Circuit:

- Since the digital circuits are highly vulnerable to switching and lightening surge therefore, proper isolation of the circuits with isolation transformer and surge protection circuit is required.

Multiplexors and S/H Circuits:

- Multiplexors and sample and hold (S/H) circuits are required for converting the analog signals to digital. The widely accepted Shannon"s sampling theorem is used for sampling the analog signal.

Anti Aliasing Filter

- The anti aliasing filters are basically low pass filters are basically low pass filters which block unwanted frequencies.

The Digital Input Output System

- This system actually gathers data and status reports of C.B. contacts status, other relay states, reset signals etc. Also the output systems generate and provide the tripping, alarm and any other control signal.

Central Processing Unit

- It is the core component of the system which performs all the logics and algorithms regarding different characteristics, maintains timing function and communicates with the external device. Therefore, this the most vital block of the numerical system.

Event Storage System, RAM, ROM, EPROM:

- The RAM store the input sample data temporarily and buffer data permanently. It is processed during the execution of relay algorithm. The ROM stores the relay algorithm permanently. EPROM is used to store certain parameters such as the relay setting, or any other relevant data. The event storage system basically stores the historical data such as fault related data, transient data, event time data.

Communication Peripherals:

- The relay setting, data uploading and event data recording are done through various communication peripherals following a protocol IEC61850, which increases inner coordination between the relays among the local and remote substation equipment.

Power Supply

- It is powered by local station battery provided with charger.

3.3 SIGNAL CONDITIONING SUBSYSTEM

3.3.1 Transducers

- Primary power system currents and voltages are usually relatively high. Before it is possible to bring these signals to protective relays, they must therefore be reduced to much lower levels. Conventionally, currents are reduced either to 5 A or 1 A and voltages are reduced to 110 V or 120 V. This is normally achieved by using primary current and voltage transducers (CTs and VTs).

- In digital relays, however, current and voltage magnitudes are both further reduced using auxiliary transducers and/or mimic impedances within the relays to suit the requirements of the components used. Ideally the current transformer would reproduce a perfectly scaled down version of the primary signal on its secondary side. Practical transformers reproduce the secondary current with some error, because these devices incorporate 'non-ideal' elements.

- The worst condition occurs when the iron core saturates during faults. The degree of signal distortion and the time after a fault at which it occurs is heavily dependent on the total burden connected to the primary current transducers. However, in most modern practical applications, the overall burden is such that the current signal distortion is small during the measuring period.

- In cases where this is not so, it is desirable to establish the effect on performance and, if necessary, incorporate means within the relay software to ensure that integrity of measurement is maintained. Some work on compensating for current transducer saturation is to be found in the literature, although, in most applications, this is unnecessary. In what follows, the effect of current transducer saturation will, for brevity, be assumed negligible.

- Electromagnetic VTs generally produce a very high-fidelity voltage signal, though in fact these are rarely used at system voltages above typically 100 kV. At higher voltages, the use of capacitor voltage transformers (CVTs) is commonplace.

- Unfortunately, the transient response of CVTs varies widely according to the type of transducer involved and the nature of the total connected burden. In high-

speed relaying applications in particular, account needs to be taken of the rather poor fidelity of the voltage signals that emanate from CVTs, there being many examples in the literature by which the digital algorithm compensates for such effects. A detailed consideration of such techniques is not necessary in a text of this type.

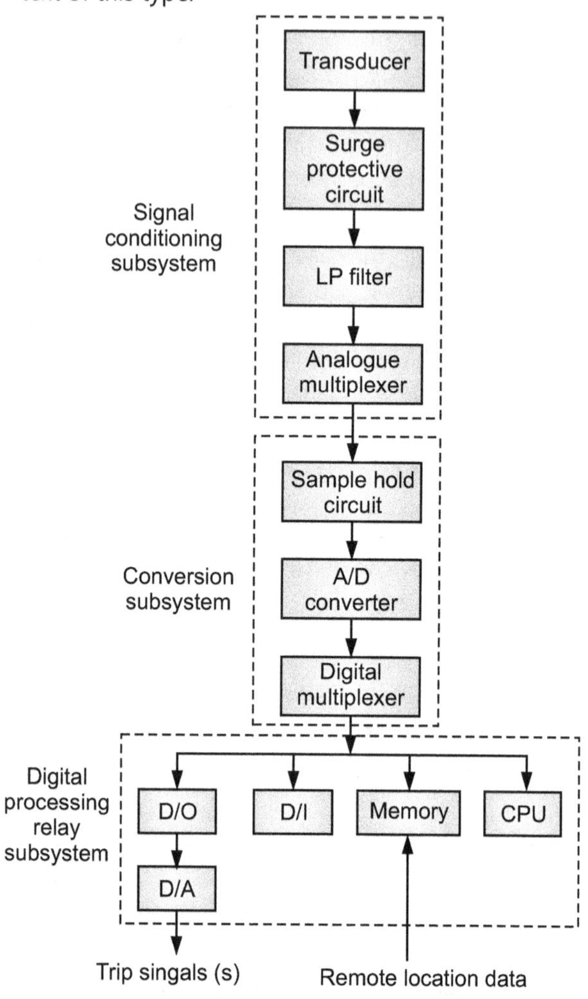

Fig. 3.1 : Basic components of a digital relay

LP = low pass

A/D = analogue-to-digital

D/A = digital-to-analogue

CPU = central processor unit

D/I = data input

D/O = data output

3.3.2 Surge Protection Circuits

- The current and voltage from the secondaries of the CTs and VTs is connected to surge protective circuits, which typically consist of capacitors and isolating transformers (Fig. 3.2). Zener diodes are also commonly used to protect electronic circuits against surges, though their placement depends on the exact physical circuit arrangement used.

- In practice, it is common to convert the secondary current measurands into low-level voltage signals by means of a suitably connected burden and/or current-to-voltage amplifier arrangement. The latter normally use careful screening techniques and are often accommodated inside separately screened self-contained modules separated from the digital signal processing hardware.

3.3.3 Analogue Filtering

- It is normally necessary to perform analogue filtering of the signals received from the CTs and PTs. In practice, the amount of filtering depends on the data requirements of a particular digital relay. Such filtering is usually performed using low-pass filters to remove unwanted high frequencies before sampling.

- The sampling theorem requires that analogue signal components above a certain frequency (which in turn is related to the digital sampling frequency) be attenuated to avoid errors in subsequent digital processing.

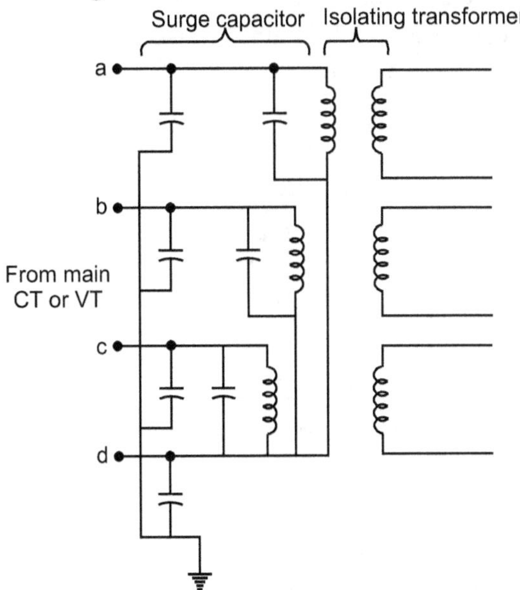

Fig. 3.2 : Simple surge protective circuit

- It is this 'anti-aliasing' function that is importantly fulfilled by the analogue low-pass filters, which must be designed with a cutoff frequency *(fc)* that performs satisfactory signal component rejection above a given frequency. Fig. 3.3(a) shows the characteristics of an ideal low-pass filter, which transfers signal components of frequencies below the cutoff frequency with zero attenuation, while components above the cutoff frequency are attenuated to zero.

- The effect of introducing a practical low-pass filter is shown in Fig. 3.3(b), from which it can be seen that in practice it is not possible to achieve such a pronounced transition between the pass and stop bands. The dynamic characteristic of the low-pass filters, as well as their steady-state characteristics, are important.

- Among the more important factors are

  - ➤ The rise time, which gives an indication of how long it takes the output of a low-pass filter to traverse its final value following a step input

  - ➤ The overshoot, which indicates by how much the filter output will exceed its steady-state value on the initial response to a unit step input

  - ➤ The settling time, which is an indication of how long it takes a given filter to settle at its steady-state output value.

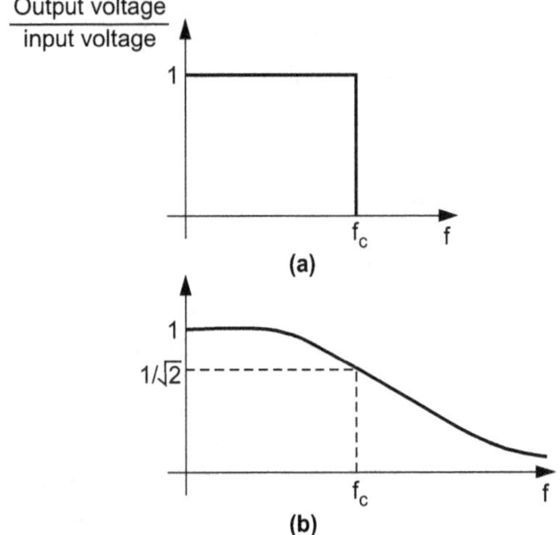

Fig. 3.3 : Characteristic of low-pass filters; (a) ideal filter response, (b) practical filter Response

- All the above features play a part in the overall dynamic response of digital relay systems. In particular, in systems where a very high speed decision is required, it is particularly important to ensure that the low-pass filter is designed to have a cutoff frequency that gives an overall performance that is not degraded by long filter delays .

3.3.4   Analogue Multiplexers

- In digital relaying applications, it is usually necessary to use an analogue multiplexer. The concept of multiplexing has its origins in communications engineering. An analogue multiplexer is a device that selects a signal from one of a number of input channels and transfers it to its output channel, thereby permitting the transmission of several signals in a serial manner over a single communication channel. The principles of multiplexing are thus as shown in Fig. 3.4, in which the solid-state multiplexer is likened to a multi-terminal rotary switch

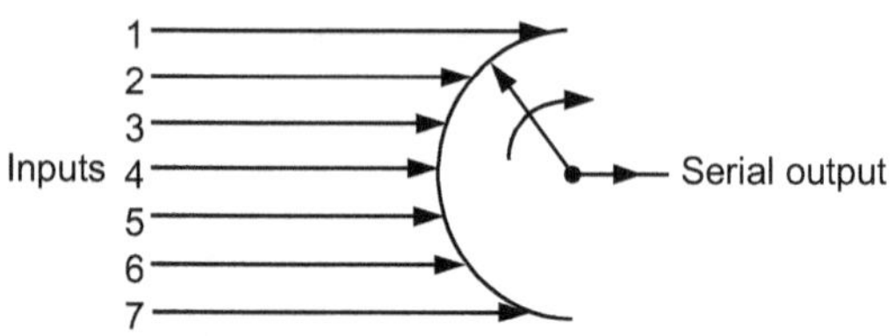

Fig. 3.4 : Principle of multiplexing

3.4 DIGITAL RELAY SUBSYSTEM

- The digital relay subsystem comprises both hardware and software. The hardware largely consists of a central processor unit (CPU), memory, data input and output (I/O). The software is influenced by two major factors. The first of these is the operating principles and performance required, which either leads to the development of a special algorithm or the implementation of an existing one.

- This factor greatly affects the determination of the sampling frequency, type of hardware structure and the data input hardware system. The second factor is the digital filtering. Sub harmonics, as well as high harmonic components, can cause false tripping, failure to trip and variation in protective relay performance. The operating principle and digital filtering must in general provide for a wide application range and requirements relating to speed of response to system faults, the latter being influenced by the necessary computation time.

- Fig. 3.5 shows an example of a flow-chart for the software of a typical digital protective relay. The algorithms used and the software required vary significantly according to the application, and much of the work of later chapters will be concerned with specific algorithms for meeting a variety of protection performance and application requirements.

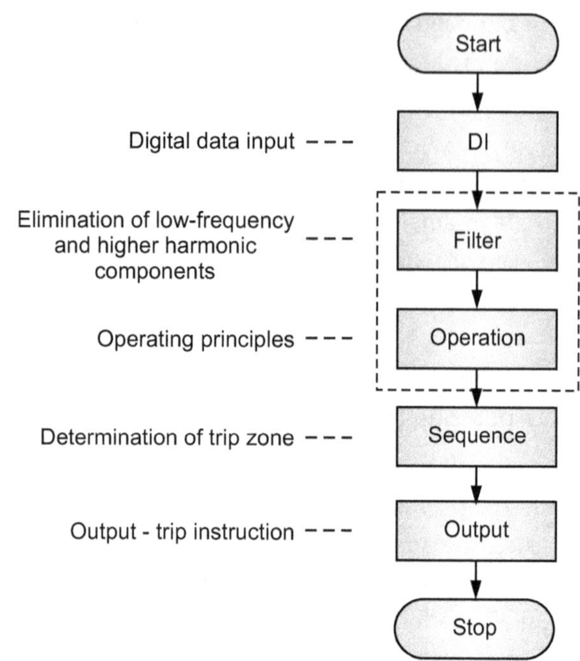

Fig. 3.5 : Flow chart for the software of a digital protective relay

3.5 THE DIGITAL RELAY AS A UNIT

- There is an inevitable tendency for relay manufacturers to develop standardized hardware, which can be used in conjunction with suitably developed software to meet a variety of production requirements and applications. Fig. 3.6 shows a block diagram of a standardised digital relay unit, and Table 3.1 gives a typical basic specification of such a unit.

Table 3.1

Data Acquisition		Sampling Frequency Data Length	720 600 sample/s 12 bit
Data processing	CPU	Device word length language	bipolar 16 bit assembler
Memory		IC memory	
Tap setting		Nonvolatile IC memory (up to 256 words)	

Table 3.1 Specification of a typical standardised digital relay unit

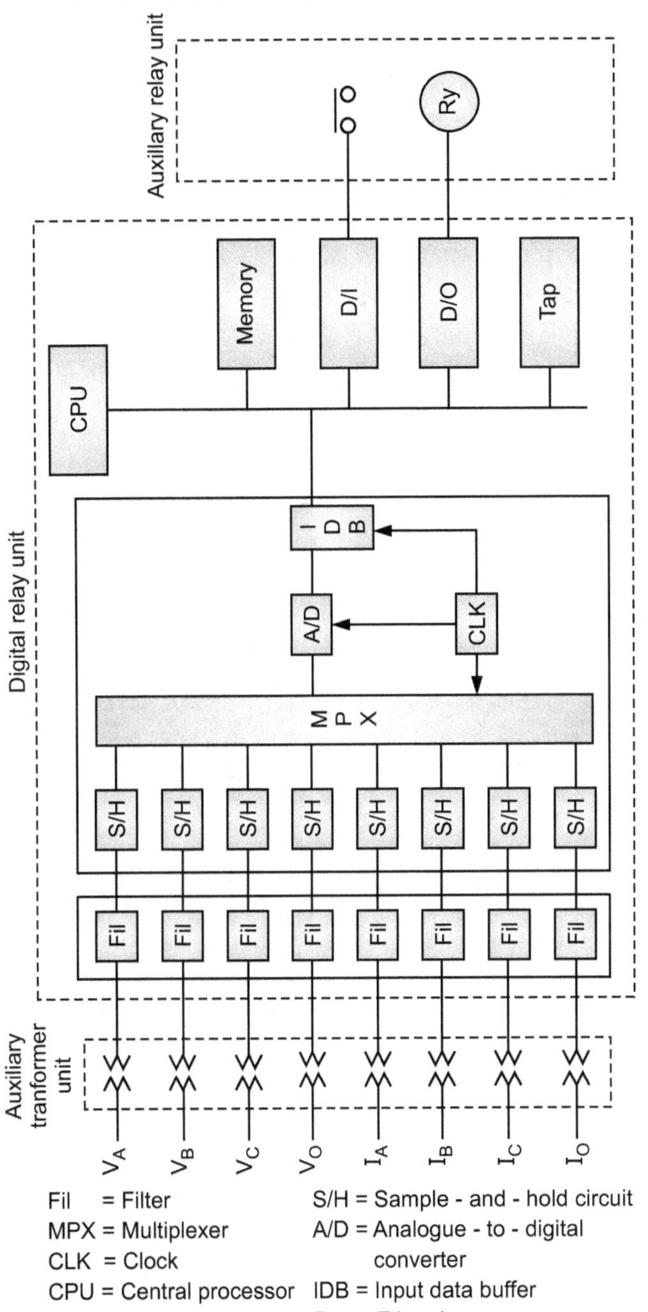

Fil = Filter
MPX = Multiplexer
CLK = Clock
CPU = Central processor unit
S/H = Sample - and - hold circuit
A/D = Analogue - to - digital converter
IDB = Input data buffer
Ry = Trip relay

Fig. 3.6 : Digital relay unit

3.6 INTRODUCTION TO NUMERICAL RELAY

- Protective relays, which started out as meters with contacts, have undergone tremendous evolution over the years. They were soon replaced by electromechanical relays which were sensitive and accurate. When vacuum tubes were in vogue, protection engineers implemented relays using vacuum tubes. Within a year of invention of the transistor, its use in protective relays was reported. With the development of large-scale integrated circuits, these were extensively used in the protective relays.

- The microprocessor that was invented around 1971 revolutionized the electronics scene in its entirety and the development of a microprocessor-based relay followed soon thereafter. However, a subtle shift in the paradigm takes place when we move on to the microprocessor-based relay, which works on numbers representing instantaneous values of the signals. Hence, the name numerical relay. Other popular nomenclatures for such relays are digital relay, computer-based relay or microprocessor-base relay.

- In numerical relays, there is an additional entity, the software, which runs in the background and which actually runs the relay. With the advent of numerical relays, the emphasis has shifted from hardware to software. Hardware is more or less the same between any two numerical relays. What distinguishes one numerical relay from the other is the software.

- The conventional non-numerical relays are go-no-go devices. They perform comparison rather than straight numerical computation. In fact, the conventional relay cleverly bypasses the problem of computation by performing comparison.

- The numerical relay does not have any such limitation because of its ability to perform real-time computation. Thus, the relay engineer need not merely implement the old relaying concepts but can devise entirely new computation-based concepts.

- The modern numerical relay has thus evolved from a torque balancing device to a programmable information processor.

- We can implement an existing relaying concept using the numerical technique. However, the possibilities of developing a new numerical relay are almost endless and there is very little standardization. The process of development of a new numerical relay is shown in the flowchart of Fig. 3.7.

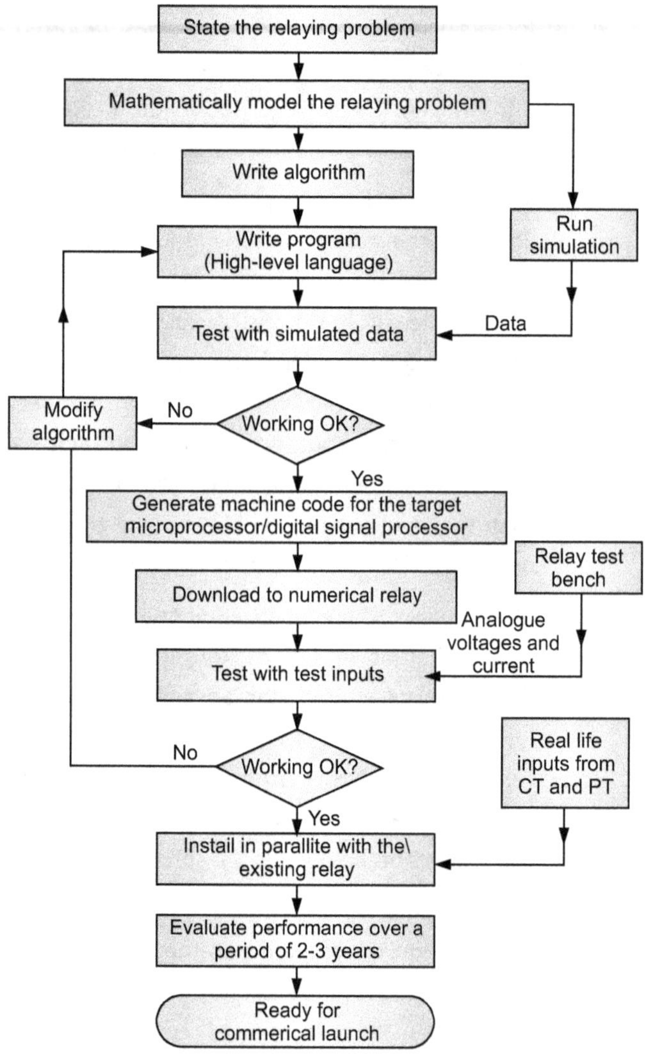

Fig. 3.7: Development cycle of a new numerical relay

3.7 BLOCK DIAGRAM OF NUMERICAL RELAY

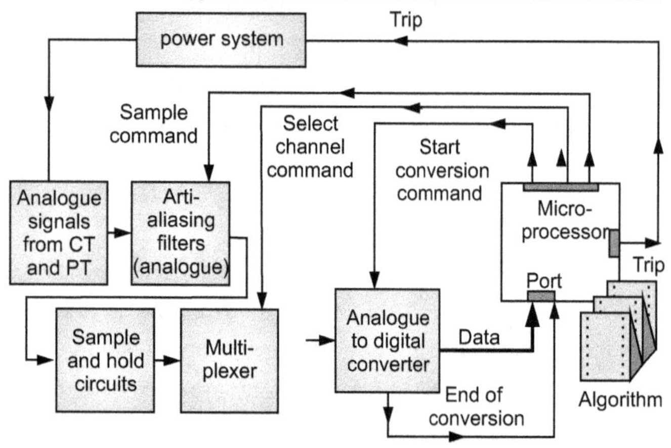

Fig. 3.8: Block diagram of numerical relay

Fig. 3.8 shows the block diagram of a numerical relay. The signals from the CTs and PTs cannot be sampled directly and converted to the digital form. This is to make sure that the signal does not contain frequency components having a frequency greater than one half of the sampling frequency. This limit is enforced by the Shannon's sampling theorem.

Comparison of maximum allowable frequency with and without S/H.

Sr. No.	Without S/H Circuit	With S/H Circuit
1.	$\dfrac{dv}{dt_{max}} = \dfrac{V_{full\ scale}}{2^n\, T_{Ad\ Coav.}}$ where $T_{AD\ conv.}$ is the conversion time of the ADC.	$\dfrac{dv}{dt_{max}} = \dfrac{V_{full\ scale}}{2^n\, T_{S/H\ aperture}}$ where $T_{S/H}$ aperture is the acquisition time of the S/H circuit.
2.	$f_{max} = \dfrac{1}{2p2^n T_{ADC\ Conv.}}$ n = ADC word length = 16 bits $T_{AD\ Conv.}$ = 10 µs (Typical) $V_m = V_{full\ scale}$ Gives f_{max} = 0.24 Hz Thus, without S/H, the ADC can handle only extremely low frequencies.	$f_{max} = \dfrac{1}{2p2^n T_{S/H\ specture.}}$ n = ADC word length = 16 bits $T_{S/H\ aperture}$ = 250 ps (Typical) $V_m = V_{full\ scale}$ Gives f_{max} = 9.7 Hz With S/H, the same ADC can not handle much higher frequencies.

- The sampled and held value is passed on to the ADC through a multiplexer so as to accommodate a large number of input signals The sample and hold circuit and the ADC work under the control of the microprocessor and communicate with it with the help of control signals such as the end-of conversion signal issued by the ADC. The ADC passes on the digital representation of the instantaneous value of the signal to the microprocessor via an input port. The output of the ADC may be 4, 8, 12, 16, or 32 bits wide or even wider. The wider the output of the ADC, the greater its resolution.

- The incoming digital values from the ADC are stored in the RAM of the microprocessor and processed by the relay software in accordance with an underlying relaying algorithm. The microprocessor issues the trip signal on one of the bits of its output port which is then suitably processed so as to make it compatible with the trip coil of the CB. The microprocessor can also be used to communicate with other relays or another supervisory computer, if so desired. The relaying program or the relay software, which resides in the EPROM, can only be upgraded or modified by authorized personnel. Thus, new features and functionalities can be added to an existing relay by upgrading its software.

- A numerical relay can be made to run a program which periodically performs a self diagnostic test and issues an alarm signal if any discrepancy is noticed. Other features like a watch-dog timer can also be implemented, which issues an alarm if the microprocessor does not reset it, periodically, within a stipulated time of a few milliseconds. This gives an increased user confidence and improves the reliability of the relay.

3.8 NUMERICAL OVER-CURRENT PROTECTION

- Numerical over-current protection is a straightforward application of the numerical relay. Here we describe a possible method of implementing a numerical over-current relay. The algorithm first reads all the settings such as the type of characteristics to be implemented, the pick-up value I_p, the time multiplier setting in case of inverse time over-current relay or the time delay in case of DTOC relay. The algorithm of a numerical over-current relay will first extract the fundamental component of the fault current I_m, from the post-fault samples of current and establish its rms value.

- Full cycle window Fourier transforms may be used for this purpose as it effectively filters out the dc offset. It will then compare this fundamental component $I_{,,,}$ with the pick-up setting and compute the plug setting multiplier, given by (IrmsIpu)a t which the relay will be operating. Equipped with the PSM value, the relay will either compute or look up the required time delay depending upon the type of over-current characteristic that is being implemented. The algorithm will then

cause the relay to go into a wait state for a time equal to the operating time.

- At the end of this time delay, the relay will once again evaluate the rms value of the fundamental to find if the fault has already been cleared by some other relay. If the fault current still persists then the relay will issue a trip output. This signal will be suitably processed to make it compatible with the trip coil of the circuit breaker. The algorithm of a typical over-current relay is shown in Fig. 3.9.

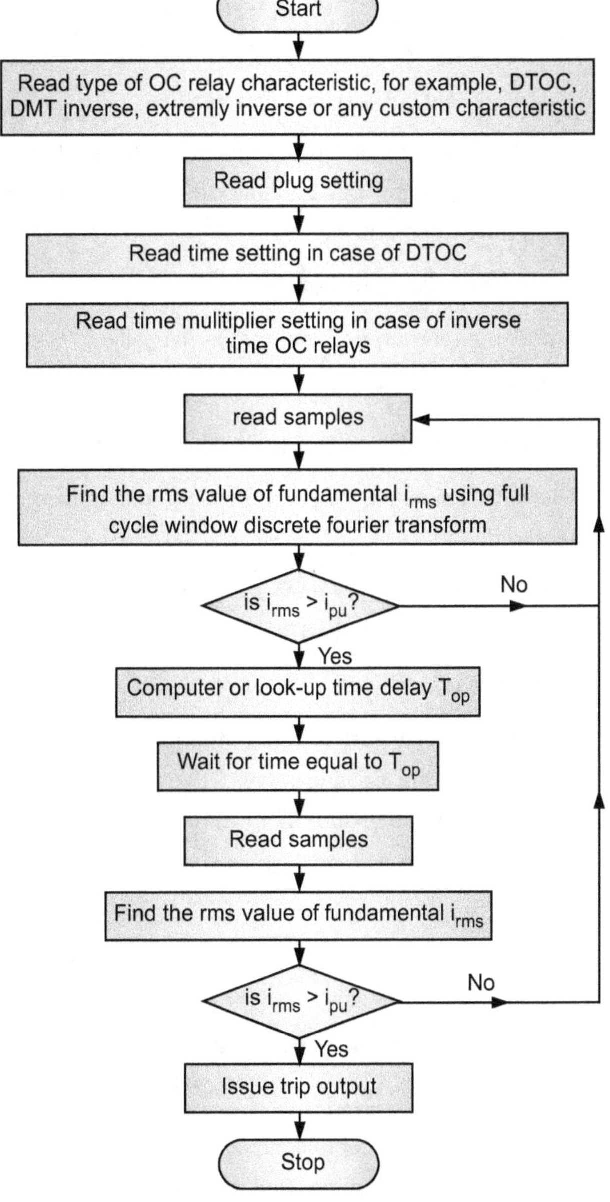

Fig. 3.9 : Flowchart for a numerical over-current relay algorithm

3.8.1 Typical Architecture of Numerical Relays

Numerical relays are made up from modules with well defined functions.

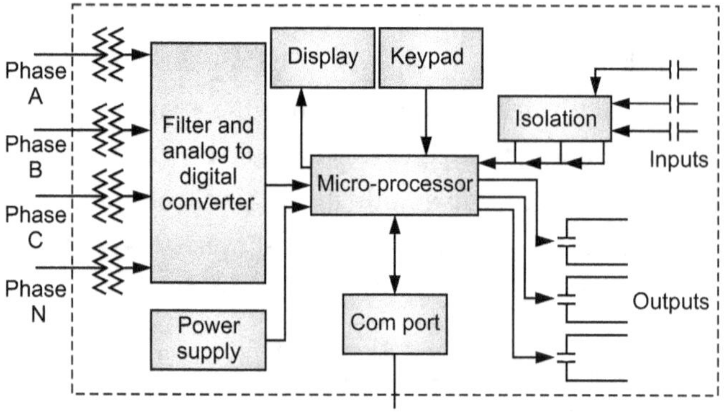

Fig. 3.10: Typical Architecture

The use of algorithm of fault diagnosis, with the help of numerical relays can be understood clearly from the following developement steps.

1. State the relaying problem.

2. Model the relaying problem mathematically

3. Write the algorithm.

4. Convert the algorithm to a high level language.

5. Test with a simulated data and modify the algorithm if required.

6. Generate the machine language code for the Micro processor/ Digital controller

7. Download it for the numerical relay

8. Test with a relay test bench. If found o.k. install it in parallel with the existing relay. Otherwise go back to step 3 to modify the algorithm and repaeat the process.

9. Evaluate with various testing for longer period and launch it commercially if found o.k in its operation after operation of 2 years independently.

3.9 NUMERICAL TRANSFORMER DIFFERENTIAL PROTECTION

Fig. 3.11 shows the conceptual block diagram for numerical protection of a transformer. The idea is to estimate the pharos value of the current on both sides of the transformer and find the phasors difference between the

two. If the magnitude of this difference is substantial, an internal fault is indicated and the trip signal should be issued. The above is a description of the simple differential scheme. All the reasons that prompt us to go in for the percentage differential relay exist here as well. Therefore, the numerical relay algorithm should be made to implement the percentage differential relay.

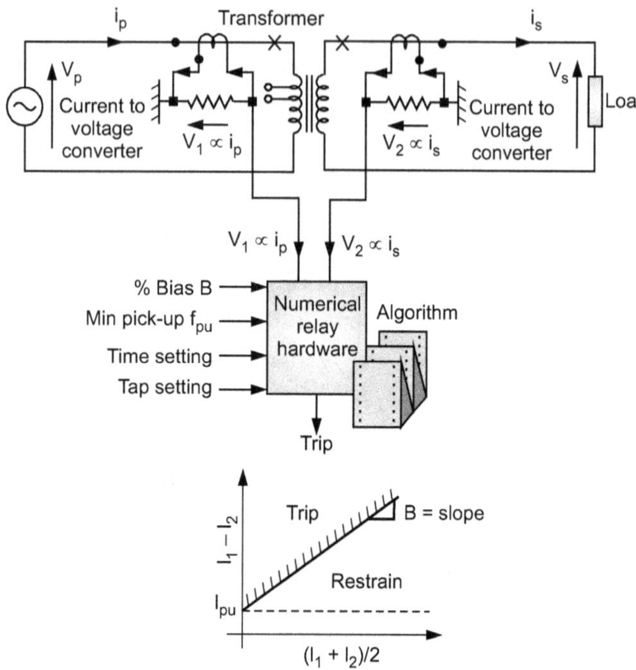

Fig. 3.11 : Block diagram of numerical protection of transformer

3.10 SAMPLING THEOREM

- The sampling theorem states that in order to preserve the information contained in a signal of frequency ω signal must be sampled at a frequency at least equal to or greater than twice the signal frequency. Thus, we must have

$$\omega_{sampling,\ min} \geq 2\omega_{signal}$$

- The lower limit on sampling frequency, equal to $2\omega_{signal}$ is known as the Nyquist limit. If the signal is sampled below the Nyquist limit, it gives rise to the phenomenon of aliasing. Aliasing is the phenomenon of the given signal being lost in the process of digitzzation, and its place being taken by a different lorver frequency waue. The above signal refers to a pure sinusoid, which contains only one frequency component to $2\omega_{signal}$.

- However, when the signal is distorted, as most real-life signals are, then the Nyquist frequency is equal to 2ω_{signal}, mar, where to 2ω_{signal} is the highest frequency component contained in the signal. Thus, the allowable sampling frequencies are those equal to or greater than to 2ω_{signal} max. Therefore, we have

$$\omega_{sampling,\,min} \geq 2\omega_{signal,\,max}$$

- The proof of the sampling theorem can be seen from Fig. 3.12.

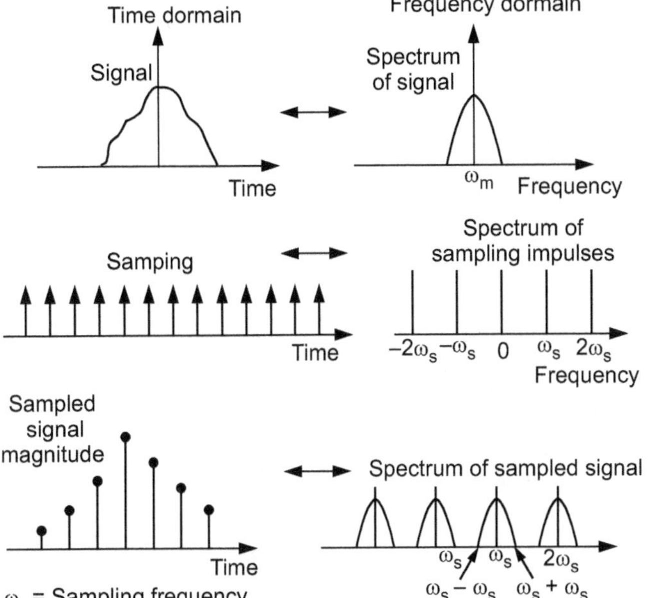

Fig. 3.12 : Proof of sampling theorem

- The signal has a frequency spectrum which extends up to ω_m. The sampling impulses, which appear at a frequency of ω_s, have a repetitive frequency spectrum which repeats at a spacing of ω_s. In the process of sampling, the sampling frequency gets modulated by the on. In order that the envelope of modulation remains distinct, the envelopes should not overlap. This will be so provided if

$$\omega_s - \omega_m > \omega_m$$

or $$\omega_s > 2\,\omega_m$$

Therefore, we have

$$\omega_{sampling,\,min} > 2\omega_{signal,\,max}$$

- If the samples of signal, which is sampled below the Nyquist rate, are fed to a digital to analogue converter (DAC), the DAC does not reproduce the original analogue signal, instead it recreates a different low frequency signal as shown in Fig. 3.13. The information contained in the original waveform is thus lost.

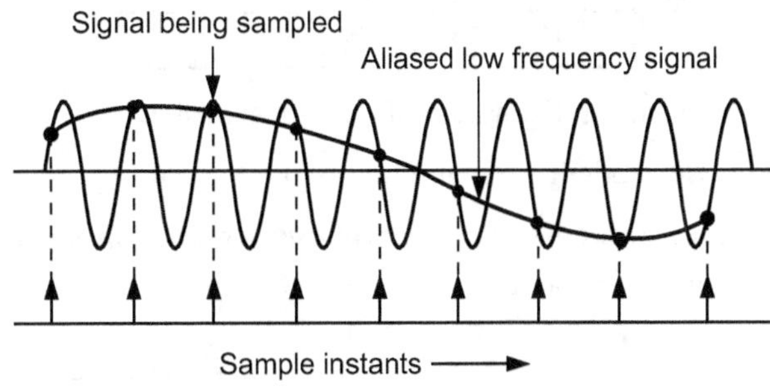

Fig. 3.13 : Phenomenon of aliasing

- Protective relaylng signals contain dc offset, a number of harmonics and random noise. whose spectrum extends from dc to several megahertz. Thus, in order to preserve information in such signals, we will need an impract~cably high sampling frequency. In order to get around this problem, the signal is first passed through a low-pass filter to shape its frequency spectrum, i.e. remove all frequencies above the cut-off frequency, so that the minimum (Nyquist) sampling rate comes down to a manageable value as shown in Fig. 3.14. The cut-off frequency is decided so that frequencies of interest are retained in the sampled version of the signal.

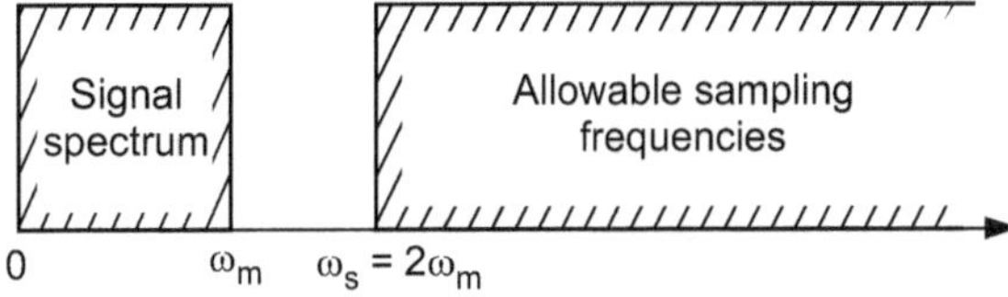

Fig. 3.14 : Minimum sampling frequency

- The above discussion would be applicable, however, if an ideal low-pass filter, which totally rejects all signals above the cut-off frequency, were available. Such filters do not exist! What we get is a finite roll-off in the stop band. Thus, the Nyquist sampling rate has to be pushed further away and a much higher minimum sampling frequency has to be used in practice as shown in Fig. 3.15.

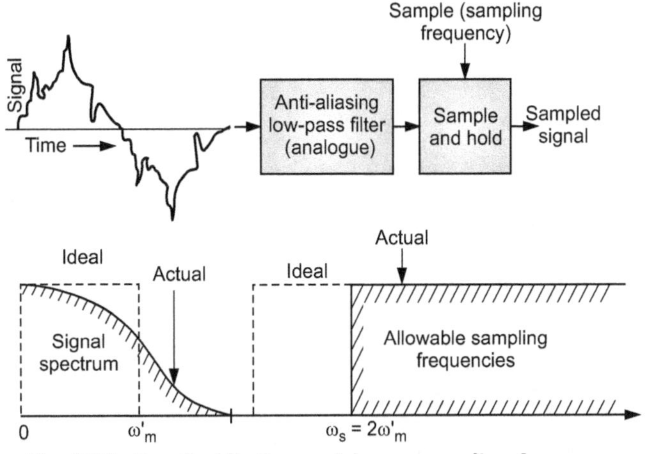

Fig. 3.15 : Practical limit on minimum sampling frequency

EXERCISE

1. Explain block diagram of numerical protection.

2. Discus briefly sampling theorem.

3. Explain numerical transformer differential protection with neat diagram.

4. Write about Basic components of a digital relay

5. List Different Components of a Digital Relay.

6. Write about Signal conditioning subsystem

7. Explain Surge Protection Circuits using digital protection

PROTECTION OF BUS-BARS AND LINES

4.1 INTRODUCTION

- Busbars and lines are important elements of electric power system and require the immediate attention of protection engineers for safeguards against the possible faults occurring on them. The methods used for the protection of generators and transformers can also be employed, with slight modifications, for the busbars and lines. The modifications are necessary to cope with the protection problems arising out of greater length of lines and a large number of circuits connected to a busbar.

- Although differential protection can be used, it becomes too expensive for longer lines due to the greater length of pilot wires required. Fortunately, less expensive methods are available which are reasonably effective in providing protection for the busbars and lines. In this chapter, we shall focus our attention on the various methods of protection of busbars and lines.

4.2 BUSBAR PROTECTION

- Busbars in the generating stations and sub-stations form important link between the incoming and outgoing circuits. If a fault occurs on a busbar, considerable damage and disruption of supply will occur unless some form of quick-acting automatic protection is provided to isolate the faulty busbar.

- The busbar zone, for the purpose of protection, includes not only the busbars themselves but also the isolating switches, circuit breakers and the associated connections. In the event of fault on any section of the busbar, all the circuit equipments connected to that section must be tripped out to give complete isolation.

- The standard of construction for busbars has been very high, with the result that bus faults are extremely rare. However, the possibility of damage and service interruption from even a rare bus fault is so great that more attention is now given to this form of protection. Improved relaying methods have been developed, reducing the possibility of incorrect operation.

4.2.1 Busbar Faults

- Majority of bus faults involve one phase and earth, but faults arise from many causes and a significant number are inter-phase clear of earth.

- With fully phase-segregated metal clad gear, only earth faults are possible ,and a protective scheme need have earth fault sensitivity only.

- For outdoor busbars, protection schemes ability to respond to inter-phase faults clear of earth is an advantage

4.2.2 Types of Protection Schemes

1. System protection used to cover bus bars
2. Frame –Earth Protection or Fault bus protection
3. Differential protection

1. System Protection

- A system protection that includes over current or distance systems will inherently give protection cover to the bus bars.

- *Over current protection* will only be applied to relatively simple distribution systems, or as a back-up protection set to give considerable time delay. *Distance protection* will provide cover with its second zone.

- In both cases, therefore ,the bus bar protection so obtained is slow

2. Frame- Earth Protection

- This is purely an earth fault system, and in principle involves simply measuring the fault current flowing from the switchgear frame to earth. To this end a current transformer is mounted on the earthing conductor and is used to energize a simple instantaneous relay.

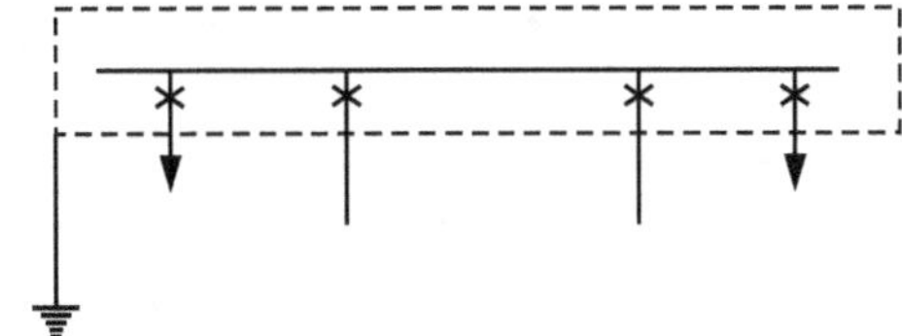

Fig. 4.1

- This protection is nothing but the method of providing earth fault protection to the bus bar assembly housed in a frame. This protection can be provided to the metal clad switchgear. The arrangement is shown in the figure below. The metal clad switchgear is lightly insulated from the earth. The enclosure of the frame housing different switchgears and bus bars is grounded through a primary of current transformer in between.

- The concrete foundation of switchgear and the other equipments are lightly insulated from the ground. The resistance of these equipments with earth is about 12 ohms. When there is an earth fault, then fault current leaks from the frame and passes through the earth connection provided. Thus the primary of C.T. senses the current due to which current passes through the sensitive earth fault relay, thereby operating the relay.

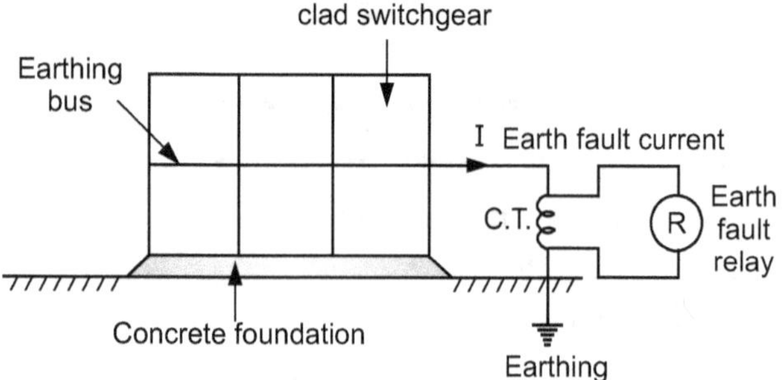

Fig. 4.2

3. Differential Protection

- The basic method for busbar protection is the differential scheme in which currents entering and leaving the bus are totalised. During normal load condition, the sum of these currents is equal to zero. When a fault occurs, the fault current upsets the balance and produces a differential current to operate a relay.

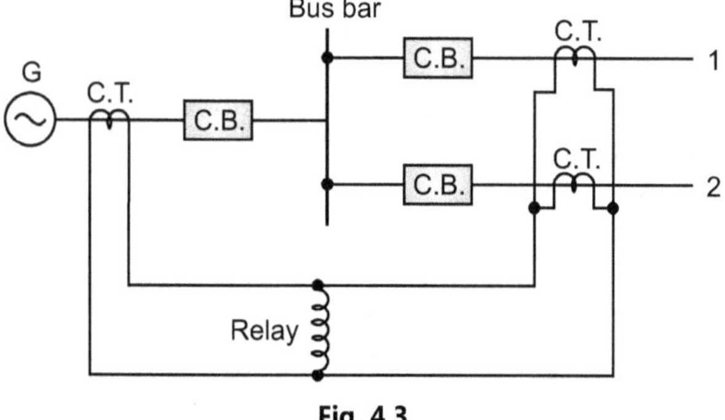

Fig. 4.3

- Fig. 4.3 shows the single line diagram of current differential scheme for a station busbar. The busbar is fed by a generator and supplies load to two lines. The secondaries of current transformers in the generator lead, in line 1 and in line 2 are all connected in parallel. The protective relay is connected across this parallel connection. All CTs must be of the same ratio in the scheme regardless of the capacities of the various circuits.

- Under normal load conditions or external fault conditions, the sum of the currents entering the bus is equal to those leaving it and no current flows through

the relay. If a fault occurs within the protected zone, the currents entering the bus will no longer be equal to those leaving it. The difference of these currents will flow through the relay and cause the opening of the generator, circuit breaker and each of the line circuit breakers.

4.3 PROTECTION OF FEEDERS

- The probability of faults occurring on the lines is much more due to their greater length and exposure to atmospheric conditions. This has called for many protective schemes which have no application to the comparatively simple cases of alternators and transformers. The requirements of line protection are :

  ➤ In the event of a short-circuit, the circuit breaker closest to the fault should open, all other circuit breakers remaining in a closed position.

  ➤ In case the nearest breaker to the fault fails to open, back-up protection should be provided by the adjacent circuit breakers.

  ➤ The relay operating time should be just as short as possible in order to preserve system stability, without unnecessary tripping of circuits. The protection of lines presents a problem quite different from the protection of station apparatus such as generators, transformers and busbars. While differential protection is ideal method for lines, it is much more expensive to use. The two ends of a line may be several kilometres apart and to compare the two currents, a costly pilot-wire circuit is required. This expense may be justified but in general less costly methods are used. The common methods of line protection are :

    1. Time-graded overcurrent protection

    2. Differential protection

    3. Distance protection

- Fig. 4.4 shows the symbols indicating the various types of relays.

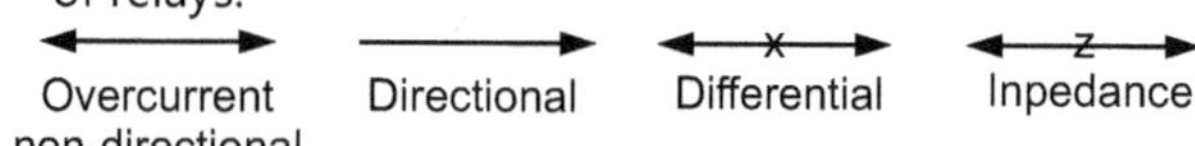

Fig. 4.4

4.4 TIME-GRADED OVERCURRENT PROTECTION

- It is customary to have two elements of over current and one element of earth fault protection system in the most elementary form of protection of three phase feeders.

- Different types of feeders employ the over current protection along with the directional relay so that proper discrimination of an internal fault is possible. Some examples are illustrated below :

1. **Radial Feeder :** The main characteristic of a radial system is that power can flow only in one direction, from generator or supply end to the load. It has the disadvantage that continuity of supply cannot be maintained at the receiving end in the event of fault. Time-graded protection of a radial feeder can be achieved by using :

   (i) Definite time relays and

   (ii) Inverse time relays.

(i) Using Definite Time Relays :

- Fig. 4.6 shows the overcurrent protection of a radial feeder by definite time relays. The time of operation of each relay is fixed and is independent of the operating current. Thus relay D has an operating time of 0.5 second while for other relays, time delay is successively increased by 0.5 second.

- If a fault occurs in the section DE, it will be cleared in 0.5 second by the relay and circuit breaker at D because all other relays have higher operating time. In this way only section DE of the system will be isolated. If the relay at D fails to trip, the relay at C will operate after a time delay of 0.5 second i.e. after 1 second from the occurrence of fault.

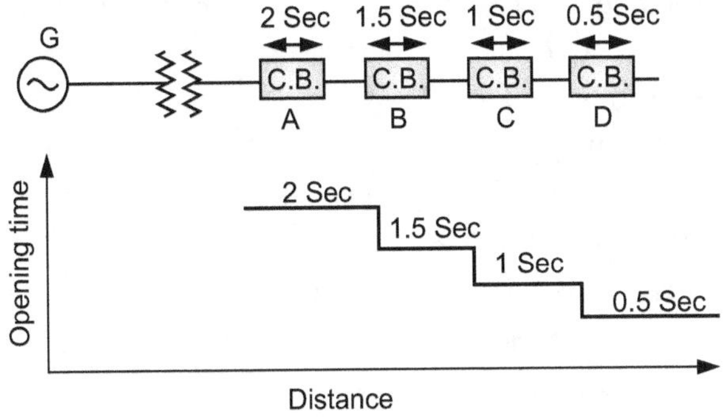

Fig. 4.5

- The disadvantage of this system is that if there are a number of feeders in series, the tripping time for faults near the supply end becomes high (2 seconds in this case). However, in most cases, it is necessary to limit the maximum tripping time to 2 seconds. This disadvantage can be overcome to a reasonable extent by using inverse-time relays.

(ii) Using Inverse Time Relays:

- Fig. 4.6 shows overcurrent protection of a radial feeder using inverse time relays in which operating time is inversely proportional to the operating current. With this arrangement, the farther the circuit breaker from the generating station, the shorter is its relay operating time.

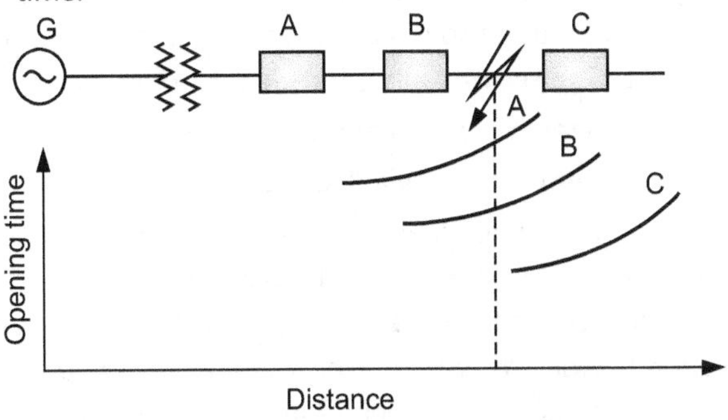

Fig. 4.6

- The three relays at *A*, *B* and *C* are assumed to have inverse-time characteristics. A fault in section *BC* will give relay times which will allow breaker at *B* to trip out before the breaker at *A*.

2. **Parallel Feeders:**

- Where continuity of supply is particularly necessary, two parallel feeders may be installed. If a fault occurs on one feeder, it can be disconnected from the system and continuity of supply can be maintained from the other feeder. The parallel feeders cannot be protected by non-directional overcurrent relays only. It is necessary to use directional relays also and to grade the time setting of relays for selective trippings.

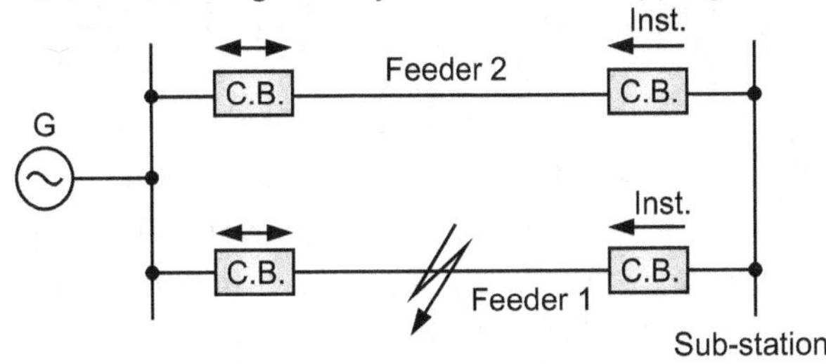

Fig. 4.7

- Fig. 4.7 shows the system where two feeders are connected in parallel between the generating station and the sub-station. The protection of this system requires that,

  ➤ Each feeder has a non-directional overcurrent relay at the generator end. These relays should have inverse-time characteristic.

  ➤ Each feeder has a reverse power or directional relay at the sub-station end. These relays should be instantaneous type and operate only when power flows in the reverse direction i.e. in the direction of arrow at P and Q. Suppose an earth fault occurs on feeder 1 as shown in Fig. 4.7. It is desired that only circuit breakers at A and P should open to clear the fault whereas feeder 2 should remain intact to maintain the continuity of supply. In fact, the above arrangement accomplishes this job. The shown fault is fed via two routes, viz.

(a) directly from feeder 1 via the relay A

(b) from feeder 2 via B, Q, sub-station and P

- Therefore, power flow in relay Q will be in normal direction but is reversed in the relay P. This causes the opening of circuit breaker at P. Also the relay A will operate while relay B remains inoperative. It is because these relays have inverse-time characteristics and current flowing in relay A is in excess of that flowing in relay B. In this way only the faulty feeder is isolated.

3. Ring Main System

- In this system, various power stations or sub-stations are interconnected by alternate routes, thus forming a closed ring. In case of damage to any section of the ring, that section may be disconnected for repairs, and power will be supplied from both ends of the ring, thereby maintaining continuity of supply.

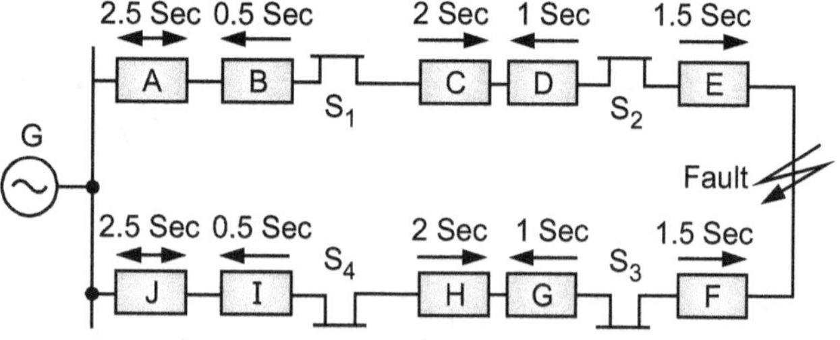

Fig. 4.8

- Fig. 4.8 shows the single line diagram of a typical ring main system consisting of one generator G supplying four sub-stations S_1, S_2, S_3 and S_4. In this arrangement, power can flow in both directions under fault conditions.

- Therefore, it is necessary to grade in both directions round the ring and also to use directional relays. In order that only faulty section of the ring is isolated under fault conditions, the types of relays and their time settings should be as follows :

  ➤ The two lines leaving the generating station should be equipped with non-directional overcurrent relays (relays at A and J in this case).

  ➤ At each sub-station, reverse power or directional relays should be placed in both incoming and outgoing lines (relays at B, C, D, E, F, G, H and I in this case).

  ➤ There should be proper relative time-setting of the relays. As an example, going round the loop $GS_1 S_2 S_3 S_4 G$; the outgoing relays (viz at A, C, E, G and I) are set with decreasing time limits e.g. A = 2.5 sec, C = 2 sec, E = 1.5 sec G = 1 sec and I = 0.5 sec Similarly, going round the loop in the opposite direction (i.e. along G S_4 S_3 S_2 S_1 G), the outgoing relays (J, H, F, D and B) are also set with a decreasing time limit e.g. J = 2.5 sec, H = 2 sec, F = 1.5 sec, D = 1 sec, B = 0.5 sec. Suppose a short circuit occurs at the point as shown in Fig. 4.8. In order to ensure selectivity, it is desired that only circuit breakers at E and F should open to clear the fault whereas other sections of the ring should be intact to maintain continuity of supply. In fact, the above arrangement accomplishes this job. The power will be fed to the fault via two routes viz (i) from G around S_1 and S_2 and (ii) from G around S_4 and S_3. It is clear that relays at A, B, C and D as well as J, I, H and G will not trip. Therefore, only relays at E and F will operate before any other relay operates because of their lower time-setting.

4.5 DIFFERENTIAL PILOT-WIRE PROTECTION

- The differential pilot-wire protection is based on the principle that under normal conditions, the current entering one end of a line is equal to that leaving the other end. As soon as a fault occurs between the two ends, this condition no longer holds and the difference of incoming and outgoing currents is arranged to flow through a relay which operates the circuit breaker to isolate the faulty line.

- There are several differential protection schemes in use for the lines. However, only the following two schemes will be discussed :

  1. Merz-Price voltage balance system

  2. Translay scheme

1. Merz-Price Voltage Balance System :

- Fig. 4.9 shows the single line diagram of Merz-Price voltage balance system for the protection of a 3-phase line. Identical current transformers are placed in each phase at both ends of the line. The pair of CTs in each line is connected in series with a relay in such a way that under normal conditions, their secondary voltages are equal and in opposition i.e. they balance each other.

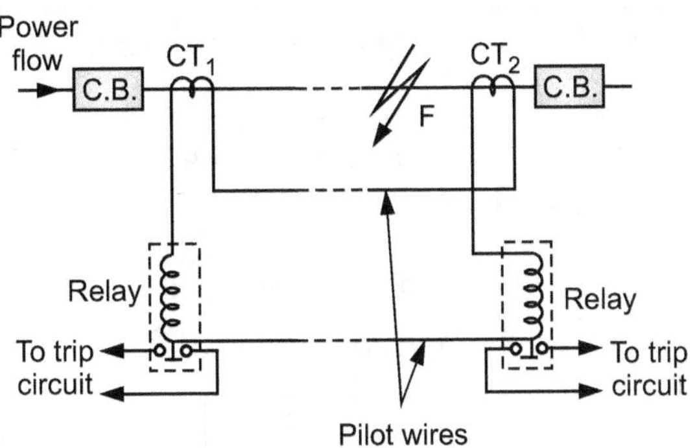

Fig. 4.9

- Under healthy conditions, current entering the line at one-end is equal to that leaving it at the other end. Therefore, equal and opposite voltages are induced in the secondaries of the CTs at the two ends of the line. The result is that no current flows through the relays. Suppose a fault occurs at point F on the line as shown in Fig. 4.9. This will cause a greater current to flow through CT1 than through CT2.

- Consequently, their secondary voltages become unequal and circulating current flows through the pilot wires and relays. The circuit breakers at both ends of the line will trip out and the faulty line will be isolated.

- Fig. 4.10 shows the connections of Merz-Price voltage balance scheme for all the three phases of the line

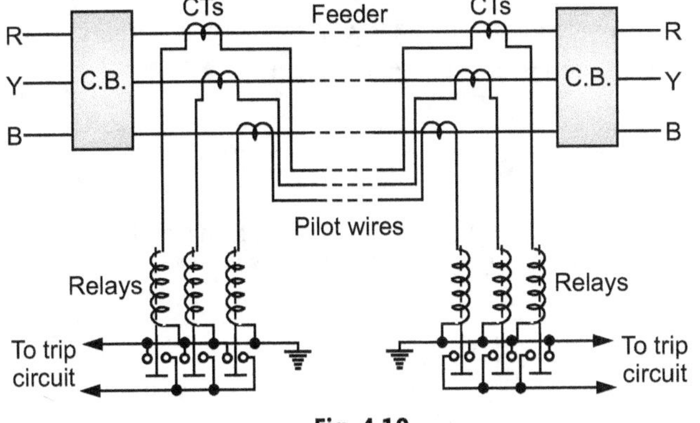

Fig. 4.10

Advantages

- This system can be used for ring mains as well as parallel feeders.

- This system provides instantaneous protection for ground faults. This decreases the possibility of these faults involving other phases.

- This system provides instantaneous relaying which reduces the amount of damage to overhead conductors resulting from arcing faults.

Disadvantages

- Accurate matching of current transformers is very essential.

- If there is a break in the pilot-wire circuit, the system will not operate.

- This system is very expensive owing to the greater length of pilot wires required.

- In case of long lines, charging current due to pilot-wire capacitance effects may be sufficient to cause relay operation even under normal conditions.

- This system cannot be used for line voltages beyond 33 kV because of constructional difficulties in matching the current transformers.

4.6 TRANSLAY SCHEME

- The translay relay is another type of differential relay. The arrangement is similar to overcurrent relay but the secondary winding is not closed on itself. Additionally copper ring or copper shading bands are provided on the central limb as shown in the figure below 4.11.

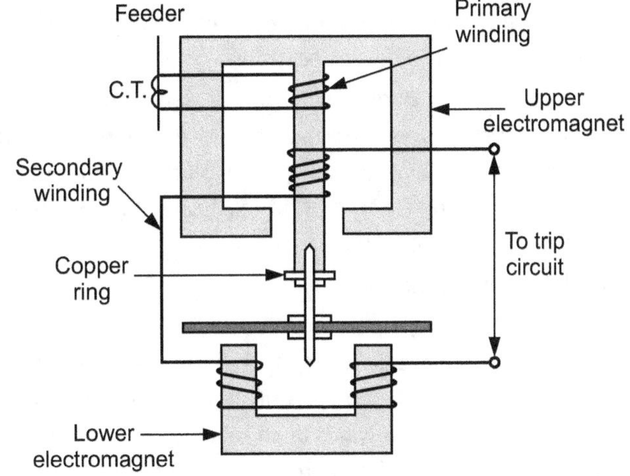

Fig. 4.11

- In this scheme, two such relays are employed at the two ends of feeder as shown in the figure below

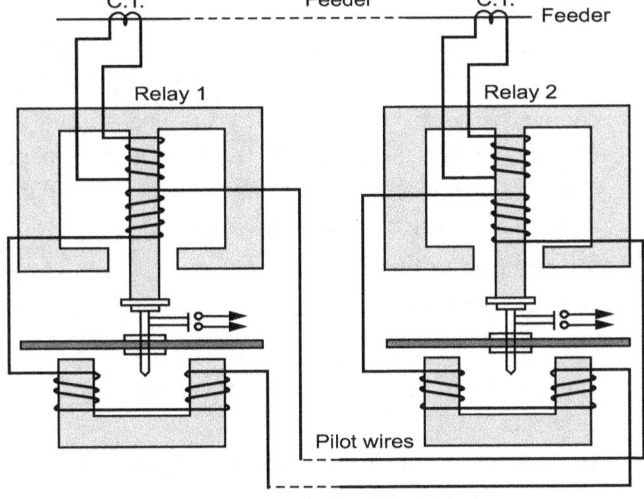

Fig. 4.12

- The secondaries of the two relays are connected to each other using pilot wires. The connection is such that the voltages induced in the two secondaries oppose each other. The copper coils are used to compensate the effect of pilot wire capacitance currents and unbalance between two currents transformers.

- Under normal operating conditions, the current at the two ends of the feeder is same. The primaries of the two relays carry the same currents inducing the same voltage in the secondaries. As these two voltages are in opposition, no current flows through the two secondaries circuits and no torque is exerted on the discs of both the relays.

- When the fault occurs, the currents at the two ends of the feeder are different. Hence unequal voltages are induced in the secondaries. Hence the circulating current flows in the secondary circuit causing torque to be exerted on the disc of each relay.

- But as the secondaries are in opposition, hence torque in one relay operates so as to close the trip circuit while in other relay the torque restricts the operation. Care must be taken so that, at least one relay operates under the fault condition.

Role of Copper Ring:

- Mainly relays may operate because of unbalance in the current transformers. The copper rings are so adjusted that the torque due to current induced in the copper ring due to primary winding of relay is restraining and do not allow the disc to rotate. It is adjusted just to neutralize the effect of unbalance existing between the current transformers.

- The copper rings also neutralize the effect of pilot capacitive currents. Though the feeder current is same at two ends, a capacitive current may flow in the pilots. This current leads the secondary voltage by 90°. The copper rings are adjusted such that no torque is exerted on the disc, due to such capacitive pilot currents. Therefore in this scheme the demerits of pilot relaying scheme is somewhat taken care of.

The Advantages of this Scheme are,

- Only two pilot wires are required.

- The cost is very low

- The current transformers with normal design can be employed.

- The capacitive effects of pilot wire currents do not affect the operation of the relays.

4.7 CARRIER CURRENT PROTECTION SYSTEM

4.7.1 The Basic Block Diagram and Various Components

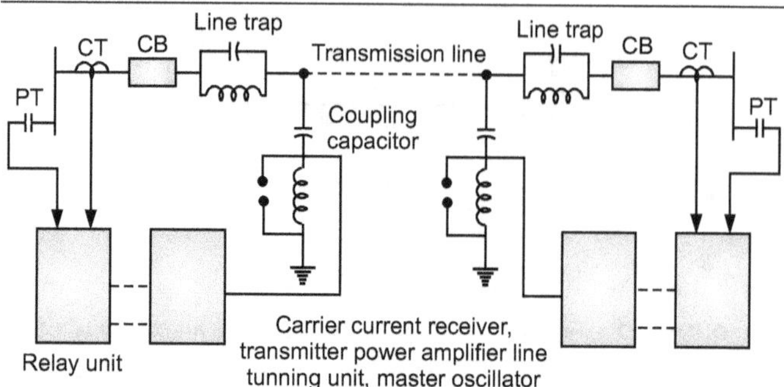

Fig. 4.13

- Schematic diagram of the carrier current scheme is shown below Fig. 4.13. Different basic components of the same are discussed below.

The Coupling Capacitor

- These coupling capacitors (CU) which offer low reactance to the higher frequency carrier signal and high reactance to the power frequency signal. Therefore, it filters out the low (power) frequency and allows the high frequency carrier waves to the carrier current equipments. A low inductance is connected to the CU, to form a resonant circuit.

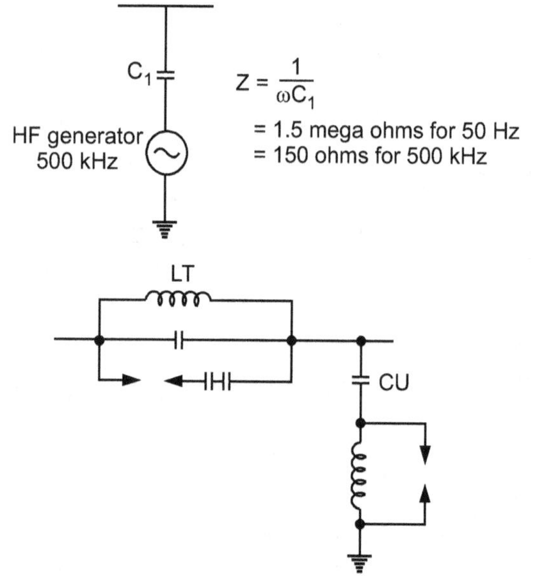

Fig. 4.14

Wave Traps

- The Wave traps (also known as Line Trap) are inserted between the busbar and the connection of the CU.

These traps are L and C elements connected in parallel, and they are tuned in such a manner that they offer low reactance to the power frequency signals and high reactance to the carrier waves.

- They ensure that neither of these different frequency signals get mixed up before being received at the bus bar. Both the CU and the Wave traps are protected from switching and lightening surges, with the help suitably designed Spark Gaps or Varistors.

Frequency Spacing

- Different frequencies are used in adjacent lines and the wave traps ensure that carrier signals of other lines do not enter a particular line section. Therefore, proper choice of frequency bands for different lines are adopted.

Transmitter Unit

- In a Transmitter unit, the carrier frequency in the range of 50 to 500 KHz of constant magnitude is generated in the oscillator, which is fed to an amplifier. Amplification is required to overcome any loss in the coupling equipments, weather conditions, Tee connections in the lines of different size and length.

- The amplifier and the oscillators are constantly energized and a connection is made between the two with the help of a control unit.

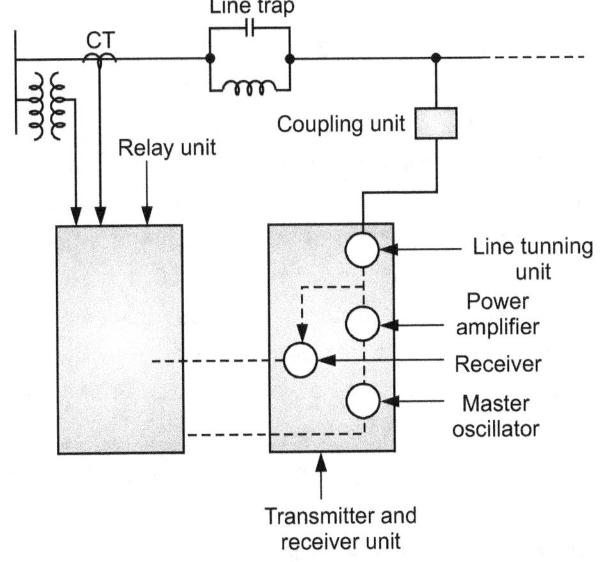

Fig. 4.15

- The Receiver unit consists of an attenuator and a Band pass filter, which restricts the acceptance of any unwanted signals. The unit also has matching

transformer to match the line impedance and that of the receiver unit.

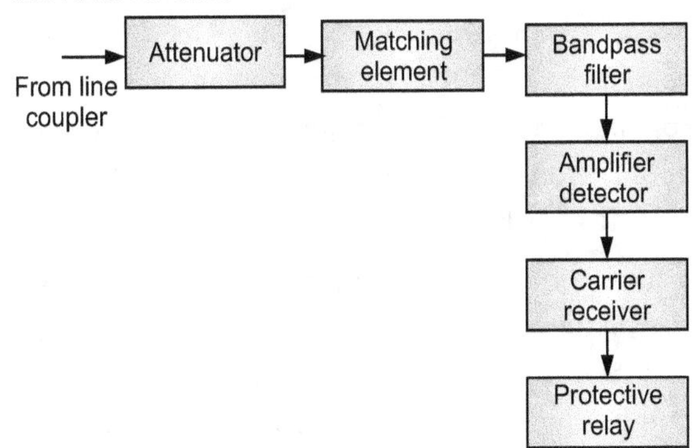

Fig. 4.16

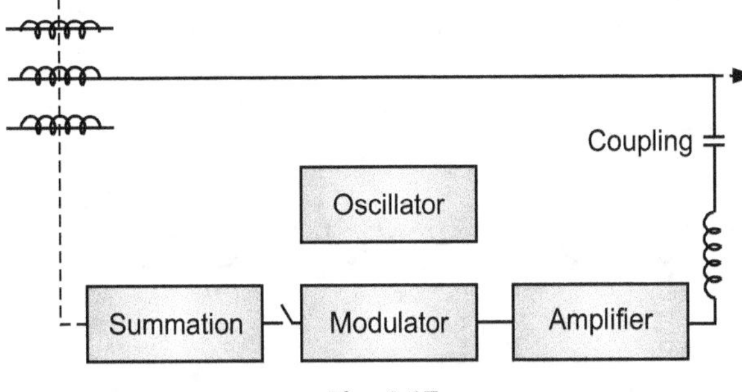

Fig. 4.17

Modulator

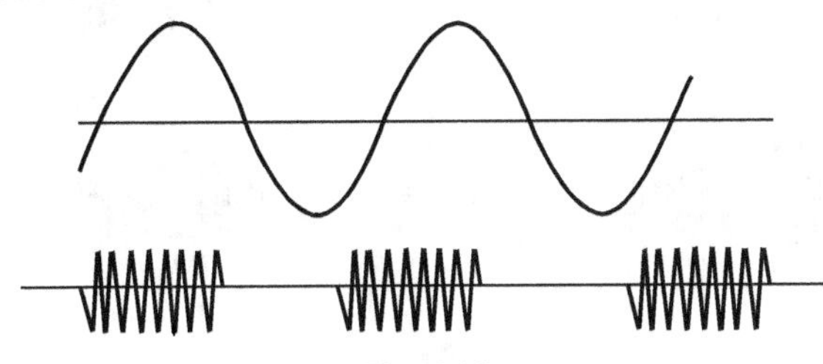

Fig. 4.18

- The Modulator modulates, the 50 Hz power signals with high frequency carrier waves and the modulated signal is fed to an amplifier. The amplifier output is transmitted via a CU. It takes half a cycle of power signal to produce requisite *Blocks of carrier* as shown above.

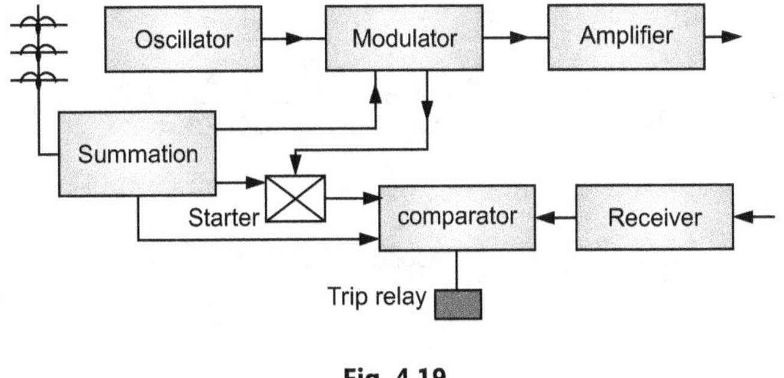

Fig. 4.19

- The CTs connected to the transmission line feed the Summation block which consist of Network sequence filters. It transforms the CT output to a single phase voltage signal that is representative of the fault condition. The voltage signal is used to control the output from the local transmitter unit, through the starting relay known as *Starter*.

- It therefore initiates comparison between the local transmitter output and the signal received from the remote receiver in the comparator. The comparator output condition then initiates the Trip relay.

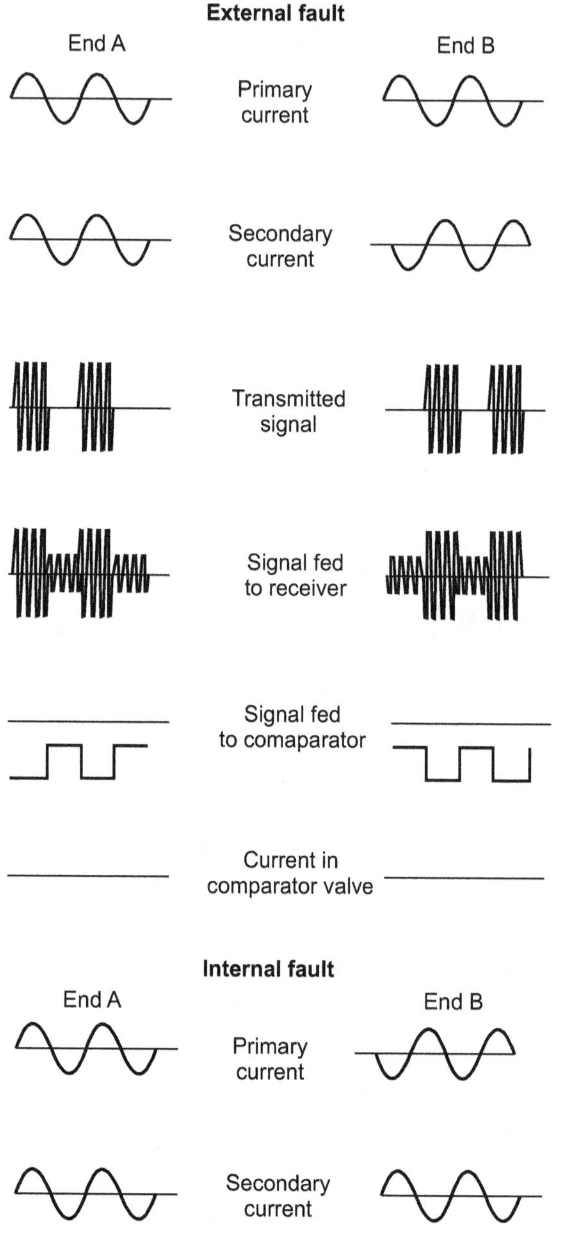

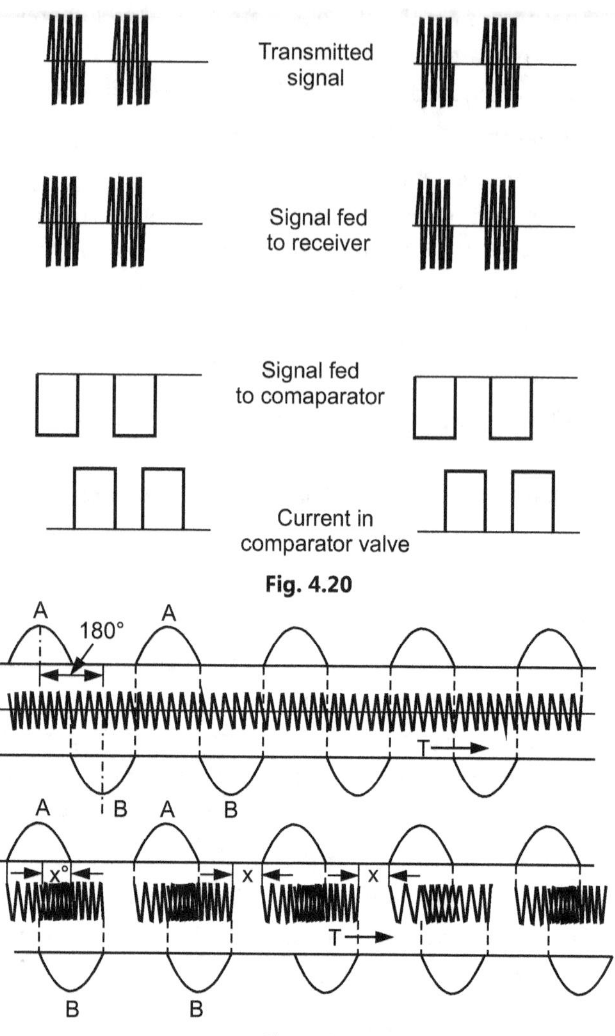

Fig. 4.20

Fig. 4.21

- The principle of Phase Comparison is one of the methods that involve decision of tripping. As shown above, the presence of blocks of carrier signals abort any tripping and its absence initiates the tripping. Therefore, in a section of transmission line, where CTs at both end buses are connected 180 degree out of phase, an absence of carrier signal can only be possible if an internal fault has occurred. However, it can be seen that such absence of carrier blocks is not possible for an external fault.

- Application advantages and multiple roles *of* CCE Pilot channel such are carrier current over the power line provides simultaneous tripping of circuit-breakers at both the ends of the line in one to three cycles. Thereby high speed fault clearing is obtained, which improves the stability of the power system. Besides there are several other merits of carrier current relaying.

There are :

- Fast, simultaneous operating of circuit-breakers at both ends.
- Auto-reclosing simultaneous reclosing signal is sent thereby simultaneous (1 to 3 cycles) reclosing of circuit breaker is obtained.
- Fast clearing prevents shocks to systems.
- Tripping due to synchronizing power surges does not occur, yet during internal fault clearing is obtained.
- For simultaneous faults, carrier current protection provides easy discrimination.
- Fast (2 cycle) and auto-reclosing circuit breakers such as air blast circuit breaker require faster relaying. Hence, the carrier current relaying is best suited for fast relaying in conjunction with modern fast circuit breaker.
- The carrier current equipment is used for several other application besides protection. They are as follows
  - ➤ Station to station communication. In power station, receiving stations and sub-stations telephones are provided. These are connected to carrier current equipment and conversion can be carried out by means of "Current Carrier Communication".
  - ➤ Control. Remote control of power station equipment by carrier signals.
  - ➤ Telemetering.

4.7.2 Media Used for Protection Signaling

- Power - line - carrier circuits
- Pilot wires

Distance Protection

- Both time-graded and pilot-wire system are not suitable for the protection of very long high voltage transmission lines. The former gives an unduly long time delay in fault clearance at the generating station end when there are more than four or five sections and the pilot-wire system becomes too expensive owing to the greater length of pilot wires required. This has led to the development of distance protection in which the action of relay depends upon the distance (or impedance) between the point where the relay is installed and the point of fault. This system provides discrimination protection without employing pilot wires.
- The principle and operation of distance relays have already been discussed. We shall now consider its application for the protection of transmission lines. Fig.

4.22 shows a simple system consisting of lines in series such that power can flow only from left to right. The relays at A, B and C are set to operate for impedance less than Z_1, Z_2 and Z_3 respectively. Suppose a fault occurs between sub-stations B and C, the fault impedance at power station and sub-station A and B will be $Z_1 + Z$ and Z respectively. It is clear that for the portion shown, only relay at B will operate. Similarly, if a fault occurs within section AB, then only relay at A will operate. In this manner, instantaneous protection can be obtained for all conditions of operation.

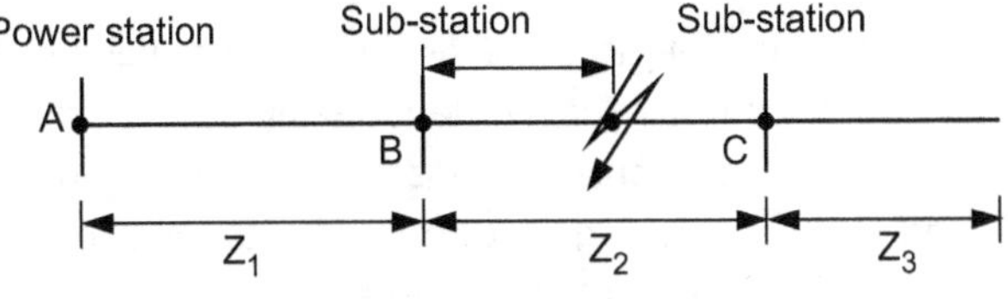

Fig. 4.22

- In actual practice, it is not possible to obtain instantaneous protection for complete length of the line due to inaccuracies in the relay elements and instrument transformers. Thus the relay at A [See Fig. 4.23 (a)] would not be very reliable in distinguishing between a fault at 99% of the distance AB and the one at 101% of distance AB. This difficulty is overcome by using 'three-zone' distance protection shown in Fig. 4.23 (b).

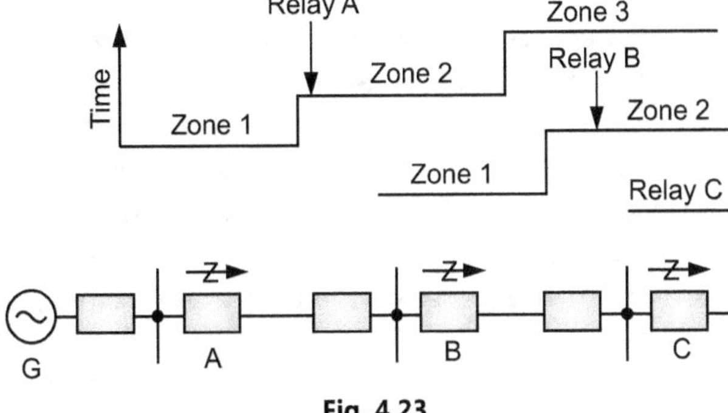

Fig. 4.23

- In this scheme of protection, three distance elements are used at each terminal. The zone 1 element covers first 90% of the line and is arranged to trip instantaneously for faults in this portion. The zone 2 element trips for faults in the remaining 10% of the line and for faults in the next line section, but a time delay is introduced to prevent the line from being tripped if the fault is in the next section. The zone 3 element provides back-up protection in the event a fault in the next section is not cleared by its breaker.

Choice between Impedance, Reactance or Mho Types

- Distance relays are used for both phase faults and ground fault protection and are preferred to overcurrent relays, latter being slow and not so selective. Distance relays are high speed relays and are also independent of changes in magnitude of the short circuit current and are not so much affected by changes in the generation capacity.

- The reactance relay is used where the arc resistance is likely to be comparable with the impedance of the protected section. The arc resistance is given by the following empirical formula

$$R_A = \frac{8750\,I}{I^{1.4}} \text{ ohms}$$

where l is the length of the arc in feed in still air and l is the fault current. The reactance relay ignores the arc resistance and hence is indispensable for short lines and for protection against ground faults. On short lines the additional potential drop in an arcing fault tends to prevent the impedance relay from operation. The disadvantage of relay, however, is that they are likely to operate undesirably on severe synchronizing power swings unless additional relay equipment is provided to prevent such operation.

- The Mho type is best suited for phase fault relaying for longer lines and particularly where severe synchronizing swings may occur. Mho relays have the advantage over two other relays of an inherent directional characteristic.

- The impedance relay is better suited for phase fault relaying for lines of moderate length than for either very short or very long lines. Arc resistance affects an impedance relay more than a reactance relay but less than a mho relay. Synchronising power swings affect an impedance relay more than a reactance relay but less than reactance relay but more than a Mho relay.

EXERCISE

1. Describe the following systems of bus-bar protection :
   (i) Differential protection
   (ii) Fault-bus protection
2. Describe the differential pilot wire method of protection of feeders.
3. Explain the Translay protection scheme for feeders.
4. Describe distance protection scheme for the protection of feeders.
5. Discuss the time-graded overcurrent protection for
   (i) Radial feeders
   (ii) Parallel feeders
   (iii) Ring main system
6. What are the requirements of protection of lines?
7. Write short-notes on the following :
   (i) Fault-bus protection
   (ii) Merz-Price voltage balance system for protection of feeders
   (iii) Translay scheme
8. Why must directional relays be used on a ring main system ?
9. How are pilot-wire relays built for transmission-line protection?
10. Do overhead systems need differential protection schemes than underground systems?
11. How do time-delay overcurrent relays work on a radial system ?
12. What Role of Copper Ring?
13. Write about Carrier Current unit protection system.

◈ ◈ ◈

PROTECTION OF ALTERNATORS AND TRANSFORMERS

5.1 INTRODUCTION

- The modern electric power system consists of several elements e.g. alternators, transformers, station bus-bars, transmission lines and other equipment. It is desirable and necessary to protect each element from a variety of fault conditions which may occur sooner or later. The protective relays can be profitably employed to detect the improper behavior of any circuit element and initiate corrective measures. As a matter of convenience, this chapter deals with the protection of alternators and transformers only.

- The most serious faults on alternators which require immediate attention are the stator winding faults. The major faults on transformers occur due to short-circuit in the transformers or their connections. The basic system used for protection against these faults is the differential relay scheme because the differential nature of measurements makes this system much more sensitive than other protective systems.

5.2 PROTECTION OF ALTERNATORS

- The generating units, especially the larger ones, are relatively few in number and higher in individual cost than most other equipments. Therefore, it is desirable and necessary to provide protection to cover the wide range of faults which may occur in the modern generating plant.

- Some of the important faults which may occur on an alternator are :

  1. Failure of prime-mover
  2. Failure of field
  3. Overcurrent
  4. Overspeed
  5. Overvoltage
  6. Unbalanced loading
  7. Stator winding faults

1. **Failure of Prime-Mover :** When input to the prime-mover fails, the alternator runs as a synchronous motor and draws some current from the supply system. This motoring conditions is known as "inverted running".

- In case of turbo-alternator sets, failure of steam supply may cause inverted running. If the steam supply is gradually restored, the alternator will pick up load without disturbing the system. If the steam failure is likely to be prolonged, the machine can be safely isolated by the control room attendant since this condition is relatively harmless. Therefore, automatic protection is not required.

- In case of hydro-generator sets, protection against inverted running is achieved by providing mechanical devices on the water-wheel. When the water flow drops to an insufficient rate to maintain the electrical output, the alternator is disconnected from the system. Therefore, in this case also electrical protection is not necessary.

- Diesel engine driven alternators, when running inverted, draw a considerable amount of power from the supply system and it is a usual practice to provide protection against motoring in order to avoid damage due to possible mechanical seizure. This is achieved by applying reverse power relays to the alternators which isolate the latter during their motoring action. It is essential that the reverse power relays have time-delay in operation in order to prevent inadvertent tripping during system disturbances caused by faulty synchronising and phase swinging.

2. **Failure of Field:** The chances of field failure of alternators are undoubtedly very rare. Even if it does occur, no immediate damage will be caused by permitting the alternator to run without a field for a short-period. It is sufficient to rely on the control room attendant to disconnect the faulty alternator manually from the system bus-bars. Therefore, it is a universal practice not to provide automatic protection against this contingency.

3. **Overcurrent:** It occurs mainly due to partial breakdown of winding insulation or due to overload on the supply system. Overcurrent protection for alternators is considered unnecessary because of the following reasons :

- The modern tendency is to design alternators with very high values of internal impedance so that they will stand a complete short-circuit at their terminals for sufficient time without serious overheating. On the occurrence of an overload, the alternators can be disconnected manually.

- The disadvantage of using overload protection for alternators is that such a protection might disconnect the alternators from the power plant bus on account of some momentary troubles outside the plant and, therefore, interfere with the continuity of electric service.

4. **Overspeed :** The chief cause of overspeed is the sudden loss of all or the major part of load on the alternator. Modern alternators are usually provided with mechanical centrifugal devices mounted on their driving shafts to trip the main valve of the prime-mover when a dangerous over speed occurs.

5. **Over-Voltage :** The field excitation system of modern alternators is so designed that overvoltage conditions at normal running speeds cannot occur. However, overvoltage in an alternator occurs when speed of the prime-mover increases due to sudden loss of the alternator load.

- In case of steam-turbine driven alternators, the control governors are very sensitive to speed variations. They exercise a continuous check on overspeed and thus prevent the occurrence of overvoltage on the generating unit. Therefore, over-voltage protection is not provided on turbo-alternator sets.

- In case of hydro-generator, the control governors are much less sensitive and an appreciable time may elapse before the rise in speed due to loss of load is checked. The over-voltage during this time may reach a value which would over-stress the stator windings and insulation breakdown may occur. It is, therefore, a usual practice to provide over-voltage protection on hydro-generator units. The over-voltage relays are operated from a voltage supply derived from the generator terminals. The relays are so arranged that when the generated voltage rises 20% above the normal value, they operate to

(i) Trip the main circuit breaker to disconnect the faulty alternator from the system

(ii) Disconnect the alternator field circuit

6. **Unbalanced Loading :** Unbalanced loading means that there are different phase currents in the alternator. Unbalanced loading arises from faults to earth or faults between phases on the circuit external to the alternator. The unbalanced currents, if allowed to persist, may either severely burn the mechanical fixings of the rotor core or damage the field winding.

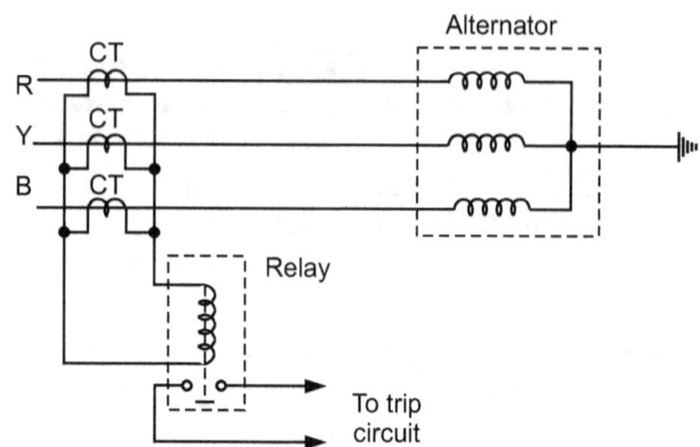

Fig. 5.1

- Fig. 5.1 shows the schematic arrangement for the protection of alternator against unbalanced loading. The scheme comprises three line current transformers, one mounted in each phase, having their secondaries connected in parallel. A relay is connected in parallel across the transformer secondaries. Under normal operating conditions, equal currents flow through the different phases of the alternator and their algebraic sum is zero.

- Therefore, the sum of the currents flowing in the secondaries is also zero and no current flows through the operating coil of the relay. However, if unbalancing occurs, the currents induced in the secondaries will be different and the resultant of these currents will flow through the relay. The operation of the relay will trip the circuit breaker to disconnect the alternator from the system.

7. **Stator Winding Faults :** These faults occur mainly due to the insulation failure of the stator windings. The main types of stator winding faults, in order of importance are:

(i) fault between phase and ground

(ii) fault between phases

(iii) inter-turn fault involving turns of the same phase winding. The stator winding faults are the most dangerous and are likely to cause considerable damage to the expensive machinery. Therefore, automatic protection is absolutely necessary to clear such faults in the quickest possible time in order to minimize the extent of damage. For protection of alternators against such faults, differential method of protection (also knows as Merz-Price system) is most commonly employed due to its greater sensitivity and reliability.

5.2.1 Biased Differential Scheme (Merz-Price Scheme) for protection of Generators

- This is most commonly used protection scheme for the alternator stator windings. The scheme is also called biased differential protection and percentage differential protection. The figure below 5.2 shows a schematic arrangement of Merz-Price protection scheme for a star connected alternator.

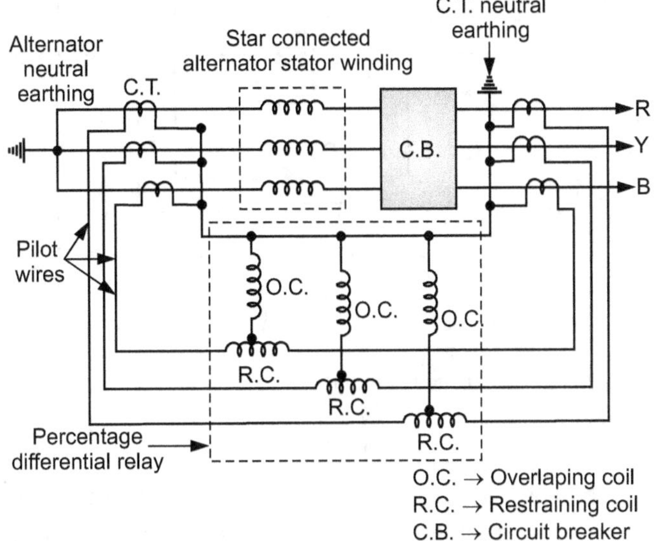

Fig. 5.2

- The differential relay gives protection against short circuit fault in the stator winding of a generator. When the neutral point of the windings is available then, the C.T.s may be connected in star on both the phase outgoing side and the neutral earth side, as shown in the above Fig. 5.2. But, if the neutral point is not available, then the phase side CTs are connected in a residual connection, so that it can be made suitable for comparing the current with the generator ground point CT secondary current.

- The restraining coils are energized from the secondary connection of C.T.s in each phase, through pilot wires. The operating coils are energized by the tappings from restraining coils and the C.T. neutral earthing connection.

- The similar arrangement is used for the delta connected alternator stator winding, as shown below Fig. 5.3.

- This scheme provides very fast protection to the stator winding against phase to phase faults and phase to ground faults. If the neutral is not grounded or grounded through resistance then additional sensitive earth fault relay should be provided.

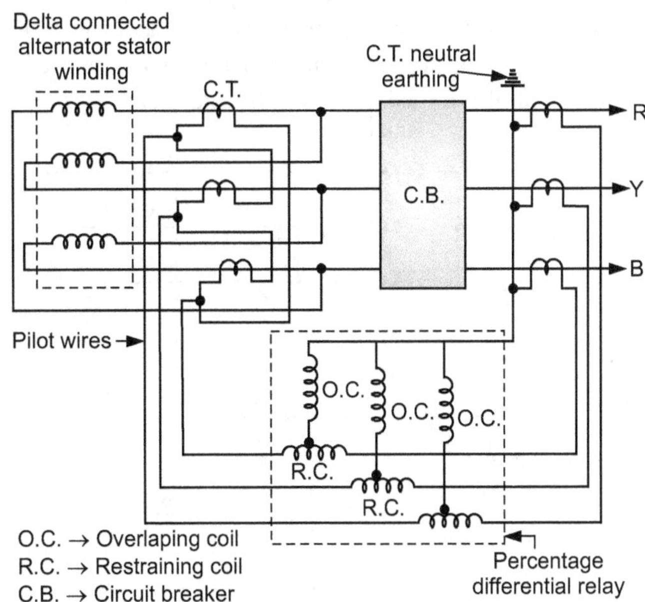

Fig. 5.3

The Advantages of this Scheme are,

- Very high speed operation with operating time of about 15 msec.

- It allows low fault setting which ensures maximum protection of machine windings.

- It ensures complete stability under most severe through and external faults.

- It does not require current transformers with air gaps or special balancing features.

5.2.2 Balanced Earth-Fault Protection

- In small-size alternators, the neutral ends of the three-phase windings are often connected internally to a single terminal. Therefore, it is not possible to use Merz-Price circulating current principle described above because there are no facilities for accommodating the necessary current transformers in the neutral connection of each phase winding. Under these circumstances, it is considered sufficient to provide protection against earth-faults only by the use of balanced earth-fault protection scheme. This scheme provides no protection against phase-to-phase faults, unless and until they develop into earth-faults, as most of them will.

- Schematic arrangement. Fig. 5.4 shows the schematic arrangement of a balanced earth-fault protection for a 3-phase alternator. It consists of three line current transformers, one mounted in each phase, having their secondaries connected in parallel with that of a single current transformer in the conductor joining the star point of the alternator to earth. A relay is connected across the transformers secondaries. The protection against earth faults is limited to the region between the neutral and the line current transformers.

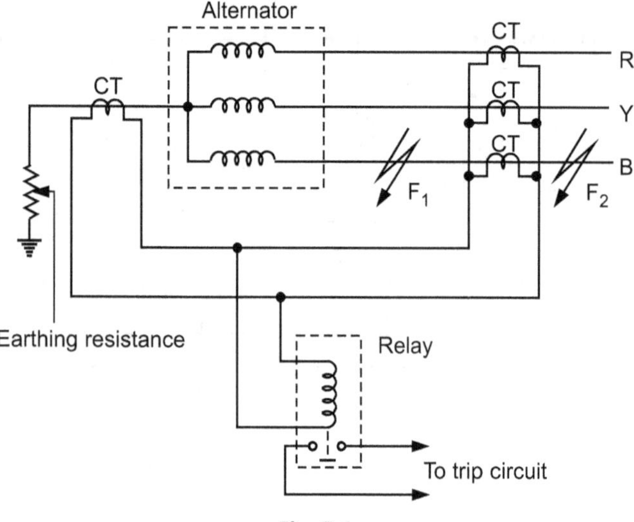

Fig. 5.4

Operation :

- Under normal operating conditions, the currents flowing in the alternator leads and hence the currents flowing in secondaries of the line current transformers add to zero and no current flows through the relay. Also under these conditions, the current in the neutral wire is zero and the secondary of neutral current transformer supplies no current to the relay.

- If an earth-fault develops at F_2 external to the protected zone, the sum of the currents at the terminals of the alternator is exactly equal to the current in the neutral connection and hence no current flows through the relay.

- When an earth-fault occurs at F_1 or within the protected zone, these currents are no longer equal and the differential current flows through the operating coil of the relay. The relay then closes its contacts to disconnect the alternator from the system.

5.2.3 Restricted Earth Fault Protection

- Generally Merz-Price protection based on circulating current principle provides the protection against internal earth faults. But for large costly generators, an additional protection scheme called restricted earth fault protection is provided. When the neutral is solidly grounded then the generator gets completely protected against earth faults. But when neutral is grounded through earth resistance, then the stator windings gets partly protected against earth faults. The

percentage of windings protected depends on the value of earthing resistance and the relay setting. In this scheme, the value of earth resistance, relay setting, current rating of earth resistance must be carefully selected.

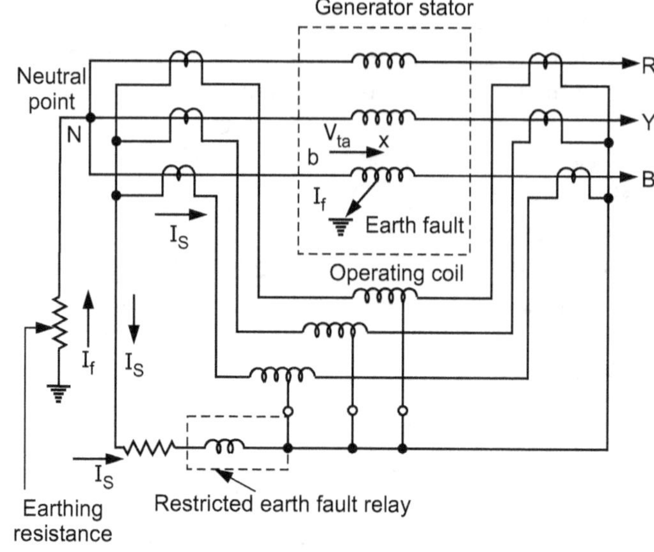

Fig. 5.5

- The earth faults are rare near the neutral point as the voltage of neutral point with respect to earth is very less. But when earth fault occurs near the neutral point, then the insufficient voltage across the fault results in a low fault current, that is less than the pickup current of relay coil. Hence the relay coil remains unprotected in this scheme. As it is able to protect a restricted portion of generator winding from earth faults, it is called a restricted earth fault protection. It is usual practice to protect 85% of the winding.

- The restricted earth fault protection scheme is shown in the above figure. Consider that earth fault occurs on phase B due to breakdown of its insulation to earth, as shown in the Fig. 5.5. The fault current If will flow through the core, frame of machine to earth and complete the path through the earthing resistance. The C.T. secondary current Is flows through the operating coil and the restricted earth fault relay coil of the differential protection. The setting of restricted earth fault relay and setting of overcurrent relay are independent of each other.

- Under this secondary current Is, the relay operates to trip the circuit breaker. The voltage V_{bx} is sufficient to drive the enough fault current If when the fault point x is away from the neutral point. If the fault point x is nearer to the neutral point then the voltage V_{bx} is small and not sufficient to drive enough fault current If. And for this If, relay can't operate. Thus part of the winding from the neutral point remains unprotected. To overcome this, if relay setting is chosen very low to make it sensitive to low fault currents, then wrong operation of relay may result.

- The relay can operate under the conditions of heavy through faults, inaccurate C.T.s, saturation of C.T.s etc. Hence practically 15% of winding from the neutral point is kept unprotected, protecting the remaining 85% of the winding against phase to earth faults.

- Let us see the effect of earth resistance on the percentage of winding which remains unprotected. Consider the earth resistance R is used to limit earth fault current. If it is very small i.e. the neutral is almost solidly grounded, then the fault current is very high. But high fault currents are not desirable hence small R is not preferred for the large machines. For low resistance R, the value of R is selected such that full load current passes through the neutral, for a full line to neutral voltage V. In medium resistance R, the earth fault current is limited to about 200A for full line to neutral voltage V, for a 60 MW machine.

- In high resistance R, the earth fault current is limited to about 10 A. This is used for distribution transformers and generator-transformer units. Now higher the value of earth resistance R, less is the earth fault current and less percentage of winding gets protected. Large percentage of winding remains unprotected.

Let   V = Full line to neutral voltage

             I = Full load current of largest capacity generator

             R = Earth resistance

The value of the resistance R is,

$$R = V/I$$

and the percentage of winding unprotected is given by,

% of winding protected = $(I_{0R}/V) \times 100$

where, Io = Maximum operating current in the primary of C.T.

- If relay setting used is 15% then I_o is 15% of full load current of the largest machine and so on.

5.2.4 The Unrestricted Earth Fault Protection

- The unrestricted earth fault protection uses a residually connected earth fault relay. It consists of three C.T.s, one in each phase. The secondary windings of three C.T.s are connected in parallel. The earth fault relay is connected across the secondaries which carries a residual current. The scheme is shown in the figure below 5.6. Where there is no fault, under normal conditions, vector sum of the three line currents is zero. Hence the vector sum of the three secondary currents is also zero.

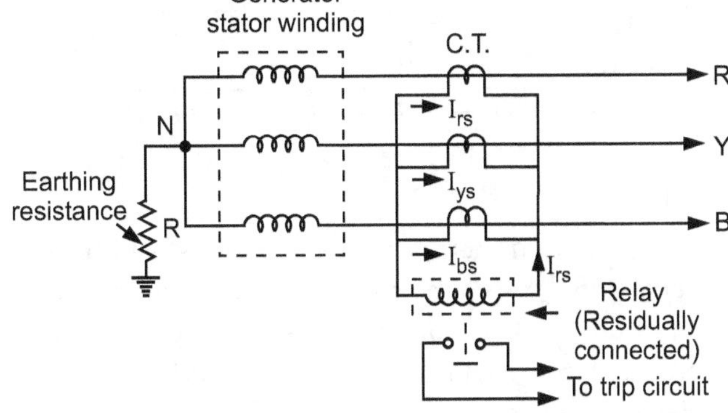

Fig. 5.6

- Currents then under normal conditions we can write,

$$\bar{I}_{rs} + \bar{I}_{ys} + \bar{I}_{bs} = 0$$

- The sum of the three currents is residual current I_{rs} which is zero under normal conditions. The earth fault relay is connected in such a way that the residual current flows through the relay operating coil. Under normal condition, residual current is zero so relay does not carry any current and is inoperative. However in presence of earth fault condition, the balance gets disturbed and the residual current I_{rs} is no more zero.

- If this current is more than the pickup value of the earth fault relay, the relay operates and opens the circuit breaker through tripping of of the trip circuit. In the scheme shown in the figure, the earth fault at any location near or away from the location of C.T.s can cause the residual current. Hence the protected zone is not definite. Such a scheme is hence called unrestricted earth fault protection.

5.2.5   Stator Inter-Turn Protection

- Merz-price circulating-current system protects against phase-to-ground and phase-to-phase faults. It does not protect against turn-to-turn fault on the same phase winding of the stator.

- It is because the current that this type of fault produces flows in a local circuit between the turns involved and does not create a difference between the currents entering and leaving the winding at its two ends where current transformers are applied. However, it is usually considered unnecessary to provide protection for inter-turn faults because they invariably develop into earth-faults.

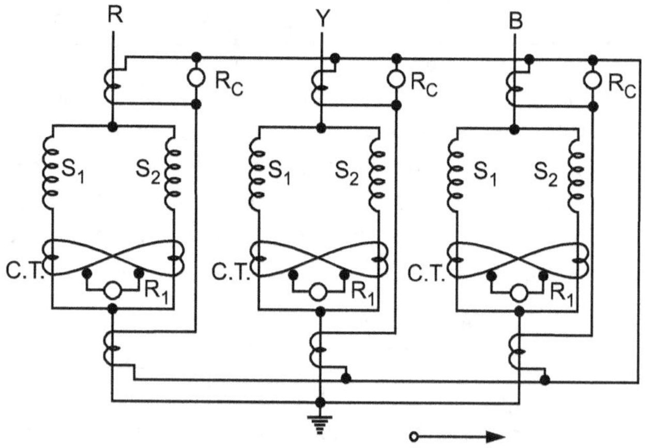

Fig. 5.7

- In single turn generator (e.g. large steam-turbine generators), there is no necessity of protection against inter-turn faults. However, inter-turn protection is provided for multi-turn generators such as hydro-electric generators.

- These generators have double-winding armatures (i.e. each phase winding is divided into two halves) owing to the very heavy currents which they have to carry. Advantage may be taken of this necessity to protect inter-turn faults on the same winding.

- Fig. 5.7 shows the schematic arrangement of circulating-current and inter-turn protection of a 3-phase double wound generator. The relays R_C provide protection against phase-to-ground and phase-to-phase faults whereas relays R_1 provide protection against inter-turn faults.

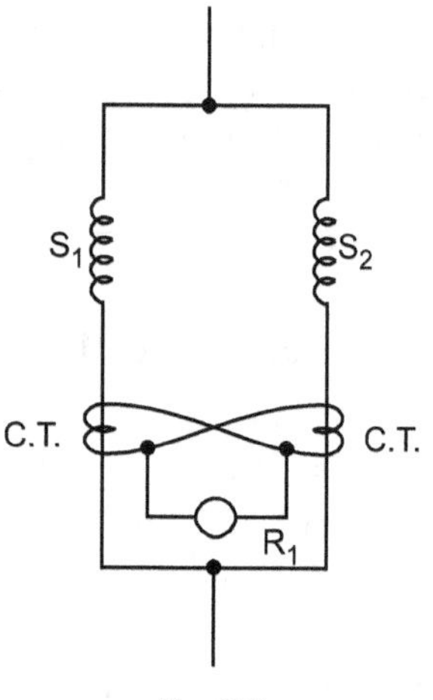

Fig. 5.8

- Fig. 5.8 shows the duplicate stator windings S_1 and S_2 of one phase only with a provision against inter-turn

faults. Two current transformers are connected on the circulating-current principle. Under normal conditions, the currents in the stator windings S_1 and S_2 are equal and so will be the currents in the secondaries of the two CTs. The secondary current round the loop then is the same at all points and no current flows through the relay R_1.

- If a short-circuit develops between adjacent turns, say on S_1, the currents in the stator windings S_1 and S_2 will no longer be equal. Therefore, unequal currents will be induced in the secondaries of CTs and the difference of these two currents flows through the relay R_1. The relay then closes its contacts to clear the generator from the system.

5.2.6 Negative Sequence Protection

- The negative sequence component can be detected by the use of a filter network. Many negative sequence filter circuits have been evolved.

- One typical negative sequence filter circuit is as follows :

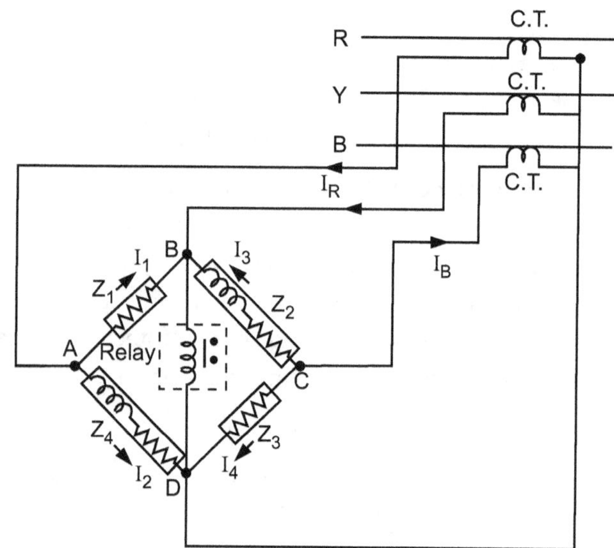

Fig. 5.9

Positive Sequence

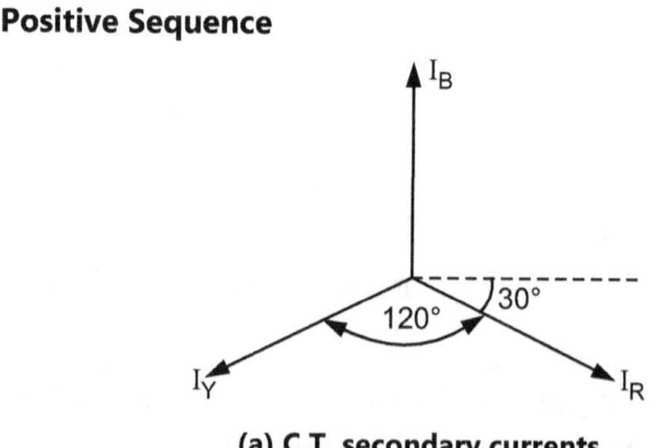

(a) C.T. secondary currents

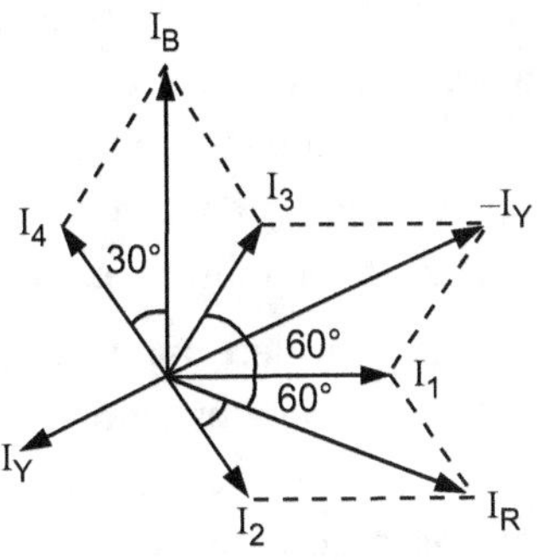

(b) Vector sum

Fig. 5.10

Negative Sequence

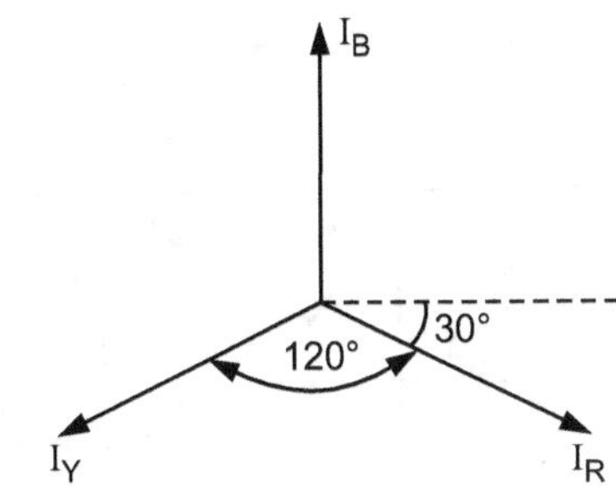

(a) C.T. secondary currents

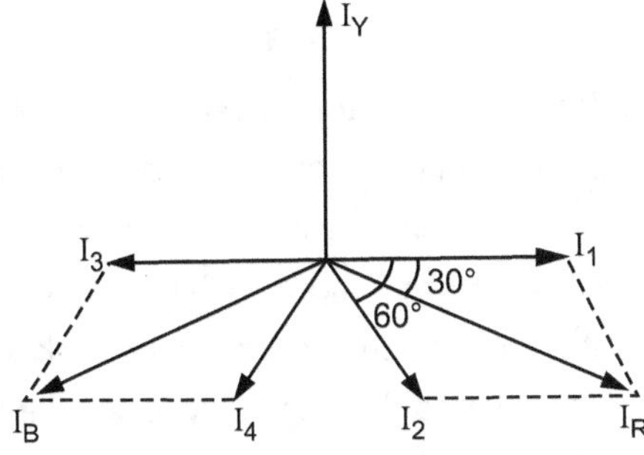

(b) Vector sum

Fig. 5.11

Zero Sequence

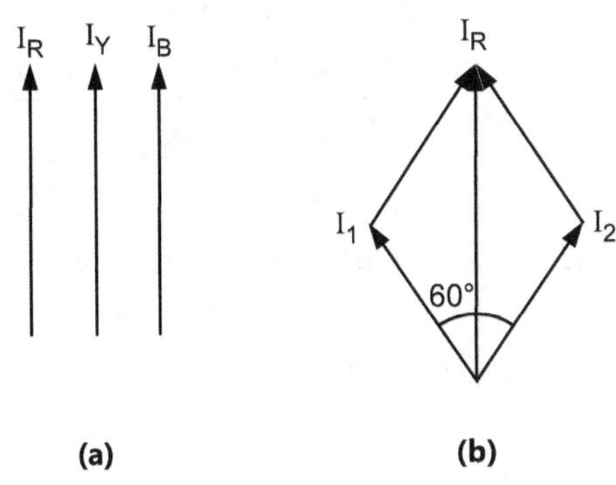

(a)   **(b)**

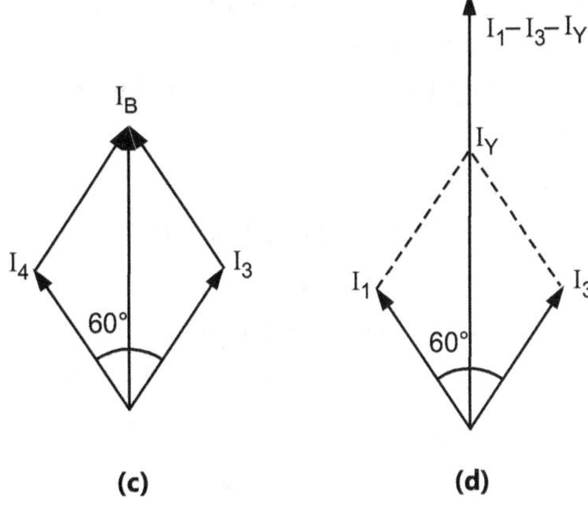

(c)   **(d)**

Fig. 5.12

- Basically it consists of a resistance bridge network as depicted in the first figure showing the circuit connection. The magnitudes of the impedances of all the branches of the network are equal. The impedances Z_1 and Z_3 are purely resistive while the impedances Z_2 and Z_4 are the combinations of resistance and reactance. The currents in the branches Z_2 and Z_4 lag by 60° from the currents in the branches Z_1 and Z_3. The vertical branch B-D basically consists of an over current element with inverse time characteristics having negligible impedance compared to the bridge impedances

5.2.7   Positive Sequence Operation

- The current I_R gets divided into two equal parts I_1 and I_2. And I_2 lags I_1 by 60°. The phasor diagram is shown in the Fig.

$$\bar{I}_1 + \bar{I}_2 = \bar{I}_{rs}$$

Let, $I_1 = I_2 = I$

- The perpendicular is drawn from point A on the diagonal meeting it at point B, as shown in the Figure. This bisects the diagonal.

$$OB = I_R/2$$

Now in triangle OAB,

$$\text{Cos } 30 = OB/OA$$

$$\sqrt{3}/2 = (I_R/2)/I$$

$$I = I_R/\sqrt{3} = I_1 = I_2 \qquad \ldots(1)$$

- Now I_1 leads I_R by 30° while I_2 lags I_R by 30°. Similarly the current I_B gets divided into two equal parts I_3 and I_4. The current I_3 lags I_4 by 60°. From equation (1) we can write,

$$I_B/\sqrt{3} = I_3 = I_4 \qquad \ldots(2)$$

- The current I_4 leads by I_B while current I_3 lags I_B by 30°. The current entering the relay at the junction point B in the Fig. is the vector sum of 3 components of currents as below.

$$I_{relay} = \bar{I}_1 + \bar{I}_3 + \bar{I}_Y$$
$$= I_Y + (I_R/\surd 3) \text{ (leads } I_R \text{ by } 30°)$$
$$+ I_B/\surd 3 \text{(lags } I_B \text{ by } 30°)$$

- The vector sum as shown in the figure, is equal to zero.

  As

$$\bar{I}_1 + \bar{I}_3 = -\bar{I}_Y$$
$$\bar{I}_1 + \bar{I}_3 + \bar{I}_Y = 0$$

- Thus the current entering the relay at point B is zero. Similarly the resultant current at junction D is also zero. Thus the relay is inoperative for a balanced system.

5.2.8 Negative Sequence Operation

- Now consider that there is unbalanced load on generator or motor due to which negative sequence currents exist. The phase sequence of C.T. secondary currents is as shown in the figure 5.11 for negative sequence . The vector diagram of I_1, I_3 and I_Y is redrawn under this condition also.

- The component I_1 and I_3 are equal and opposite to each other at the junction point B. Hence I_1 and I_3 cancel each other. Now the relay coil carries the current I_Y and when this current is more than a predetermined value, the relay trips closing the contacts of trip circuit which opens the circuit breaker.

5.2.9 Zero Sequence Operation

$$\bar{I}_R = \bar{I}_1 + \bar{I}_2 \,,$$
$$\bar{I}_B = \bar{I}_3 + \bar{I}_4 \,,$$
$$\bar{I}_1 + \bar{I}_3 = \bar{I}_Y$$

- The total current through relay is $\bar{I}_1 + \bar{I}_3 + \bar{I}_Y$. Thus under zero sequence currents the total current of twice the zero sequence current flows through the relay. Hence the relay operates to open the circuit breaker.

- To make the relay sensitive to only negative sequence currents by making it inoperative under the influence of zero sequence currents is possible by connecting the current transformers in Delta, as in that case no zero sequence current can flow in the network.

5.2.10 Loss of Excitation Protection of Generators

- The loss of excitation of the generator may result in the loss of synchronism and slightly increase in the generator speed. The machine starts behaving as an induction generator. It draws reactive power from the system which is undesirable. The loss of excitation may lead to the pole slipping condition. Hence protection against loss of excitation must be provided.

- The protection is provided using directional distance type relay with the generator terminals. When there is loss of excitation, the equivalent generator impedance varies and traces a curve as shown in the following figure. The figure shows the loss of excitation characteristics along with the relay operation characteristic, on R-X diagram.

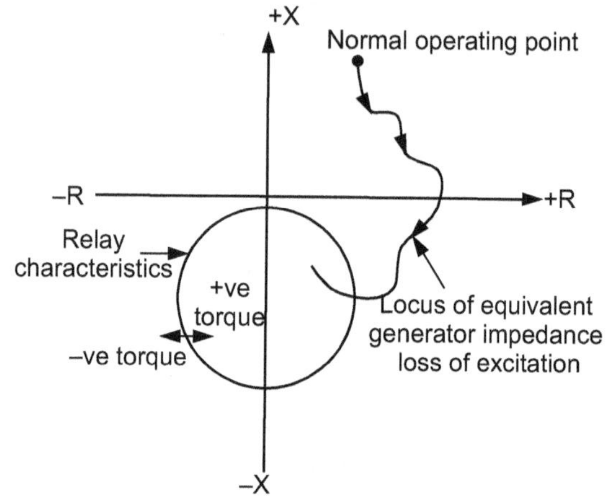

Fig. 5.13

- The equivalent generator impedance locus traces a path from first quadrant of R-X diagram to the fourth quadrant. The distance relay is used which covers the portion of the fourth quadrant where impedance locus path exists. Thus when the impedance takes value in the region covered by the relay characteristics, the relay operates.

- The relay operates when generator first starts to slip poles. Then relay trips the field circuit breaker. And it disconnects the generator from the system, too. When the excitation is regained and becomes normal, the generator can then be returned to service instantly.

SOLVED EXAMPLES

Example 5.1 : *A generator is protected by restricted earth fault protection. The generator ratings are 13.2 kV, 10 MVA. The percentage of winding protected against phase to ground fault is 85%. The relay setting is such that it trips for 20% out of balance. Calculate the resistance to be added in the neutral to ground connection.*

Solution : The given values,

$$V_L = 13.2 \text{ kV Rating}$$

$$= 10 \text{ MVA From rating,}$$

Calculate the full load current,

$$I = \frac{\text{Rating in VA}}{(\sqrt{3} \, V_L)}$$

$$= \frac{(10 \times 10_6)}{(\sqrt{3} \times 13.2 \times 10_3)}$$

$$= 437.386 \text{ A}$$

Relay setting is 20% out of balance i.e. 20% of the rated current activities the relay.

$$I_o = 4387.386 \times (20/100) = 87.477 \text{ A}$$

$$= \text{Minimum operating current}$$

$$V = \text{line to neutral voltage} = \frac{V_L}{\sqrt{3}}$$

$$= \frac{(13.2 \times 10^3)}{\sqrt{3}} = 7621.02 \text{ V}$$

% of winding unprotected = 15% as 85% is protected

$$\therefore \quad 15 = \left(\frac{RI_o}{V}\right) \times 100$$

$$= \frac{(R \times 87.477)}{7621.02} \times 100$$

$$\therefore \quad R = 13.068 \ \Omega$$

Example 5.2 : *A star connected 3 phase, 12 MVA, 1 kV alternator has a phase reactance of 10%. It is protected by Merz-Price circulating current scheme which is set to operate for fault current not less than 200 A. Calculate the value of earthing resistance to be provided in order to ensure that only 15% of the alternator winding remains unprotected.*

Solution : The given values are,

$$V_L = 11 \text{ kV}$$

$$\text{Rating} = 12 \text{ MVA}$$

$$\text{Rating} = \sqrt{3} \, V_L \, I_L$$

$$\therefore \quad 12 \times 10^6 = \sqrt{3} \times 11 \times 10^3 \times I_L$$

$$\therefore \quad I_L = \frac{(12 \times 10^6)}{(\sqrt{3} \times 11 \times 10^3)}$$

$$= 629.8366 \text{ A}$$

$$= \text{A} = \text{rated current}$$

$$V = \frac{V_L}{\sqrt{3}} = \frac{(11 \times 10^3)}{\sqrt{3}}$$

$$= 6350.8529 \text{ V}$$

$$\text{\% Reactance} = \left(\frac{X}{V}\right) \times 100$$

where X = reactance per phase

and I = rated current

$$\therefore \quad 10 = \left(\frac{629.8366X}{6350.82529}\right) \times 100$$

$$\therefore \quad X = 1.0083 \ \Omega$$

$\therefore$ Reactance of unprotected winding

$$= (\text{\% of unprotected winding}) \times (X)$$

$$= \left(\frac{15}{100}\right) \times 1.0083$$

$$= 0.1512 \ \Omega$$

$$v = \text{Voltage induced in unprotected winding}$$

$$= \left(\frac{15}{100}\right) \times V$$

$$= 0.15 \times 6350.8529$$

$$= 952.6279 \text{ V}$$

$$i = \text{Fault current}$$

$$= 200 \text{ A}$$

$$Z = \text{Impedance offered to the fault}$$

$$= \frac{v}{i} = \frac{952.6279}{200}$$

$$= 4.7631 \ \Omega \qquad \qquad \text{...(1)}$$

$$Z = r + j$$

$$(\text{reactance of unprotected winding})$$

$$Z = r + j \, (0.1512) \ \Omega$$

$$\therefore \quad |Z| = \sqrt{r^2 + 0.1512^2} \qquad \text{...(2)}$$

Equating (1) and (2),

$$4.7631 = \sqrt{r^2 + 0.1512^2}$$

$$\therefore \quad 22.6875 = r^2 + 0.02286$$

$$\therefore \quad r^2 = 22.6646$$

$$\therefore \quad r = 4.7607 \ \Omega$$

This is the earthing resistance required.

Example 5.3 : *A star-connected, 3-phase, 10 MVA, 6-6 kV alternator has a per phase reactance of 10%. It is protected by Merz-Price circulating current principle which is set to operate for fault currents not less than 175 A. Calculate the value of earthing resistance to be provided in order to ensure that only 10% of the alternator winding remains unprotected.*

Solution : Let r ohms be the earthing resistance required to leave 10% of the winding unprotected (portion NA). The whole arrangement is shown in the simplified diagram of Fig. 5.14.

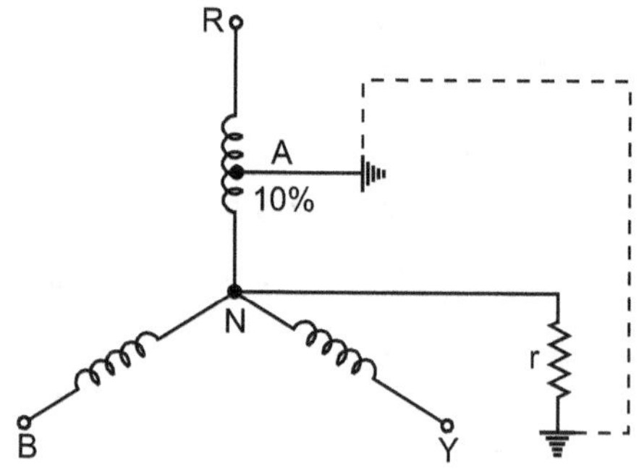

Fig. 5.14

Voltage per phase,

$$V_{ph} = \frac{6.6 \times 10^3}{\sqrt{3}} = 3810 \text{ V}$$

Full-load current,

$$I = \frac{10 \times 10^6}{\sqrt{3} \times 6.6 \times 10^3} = 875 \text{ A}$$

Let the reactance per phase be x ohms.

$$\therefore \quad 10 = \frac{\sqrt{3} \times x \times 875}{6600} \times 100$$

or $x = 0.436 \ \Omega$

Reactance of 10% winding

$$= 0.436 \times 0.1 = 0.0436 \ \Omega$$

E.M.F. induced in 10% winding

$$= V_{ph} \times 0.1 = 3810 \times 0.1$$

Impendance of offered to fault by 10% winding is

$$Z_f = \sqrt{(0.0436)^2 + r^2}$$

Earth-fault current due to 10% winding

$$= \frac{381}{Z_f} = \frac{381}{\sqrt{(0.0436)^2 + r^2}}$$

When this fault current becomes 175A, the relay will trip.

$$\therefore \quad 175 = \frac{381}{\sqrt{(0.0436)^2 + r^2}}$$

or $(0.0436)^2 + r^2 = \left(\frac{381}{175}\right)^2$

or $(0.436)^2 + r^2 = 4.715$

or $r = 2.171 \ \Omega$

Example 5.4 : *A star connected, 3-phase, 10 MVA, 6-6 kV alternator is protected by Merz-price circulating current principle using 1000/5 amperes current transformers. The star point of the alternator is earthed through a resistance of 7.5 Ω. If the minimum operating current for the relay is 0.5 A, calculate the percentage of each phase of the stator winding which is unprotected against earth-faults when the machine is operating at normal voltage.*

Solution : Let x% of the winding be unprotected.

Earthing resistance,

$$r = 7.5 \ \Omega$$

Voltage per phase,

$$V_{ph} = 6.6 \times \frac{10^3}{\sqrt{3}} = 3810 \text{ V}$$

Minimum fault current which will operate the relay

$$= \frac{1000}{5} \times 0.5 = 100 \text{ A}$$

E.M.F. induced in x% $= V_{ph} \times (x/100)$

$$= 3810 \times (x/100) = 38.1 \text{ x volts}$$

Earth fault current which x% winding will cause

$$= \frac{38.1 \text{ x}}{r}$$

$$= \frac{38.1 \text{ x}}{7.5} \text{ amperes}$$

This current must be equal to 100 A.

$$\therefore \quad 100 = \frac{38.1 \text{ x}}{7.5}$$

or   Unprotected winding,

$$x = \frac{100 \times 7.5}{38.1} = 19.69 \ \%$$

Hence 19.69% of alternator windings is left unprotected.

Example 5.5 : *A star-connected, 3-phase, 10 MVA, 6-6- kV alternator is protected by circulating current protection, the star point being earthed via a resistance r. Estimate the value of earthing resistor if 85% of the stator winding is protected against earth faults. Assume an earth fault setting of 20%. Neglect the impedance of the alternator winding.*

Solution : Since 85 % winding is to be protected, 15% would be unprotected. Let r ohms be the earthing resistance required to leave 15% of the winding unprotected.

Full-load current $= \dfrac{10 \times 10^{6}}{\sqrt{3} \times 6.6 \times 10^{3}} = 876$ A

Minimum fault current which will operate the relay

$$= 20 \% \text{ of full-load current}$$

$$= \frac{20}{100} \times 876 = 175 \text{ A}$$

Voltage induced in 15% of winding

$$= \frac{15}{100} \times \frac{6.6 \times 10^{3}}{\sqrt{3}} = 330\sqrt{3} \text{ volts}$$

Earth fault current which 15% winding will cause

$$= \frac{330\sqrt{3}}{r}$$

This current must be equal to 175 A.

$$\therefore \qquad 175 = \frac{330\sqrt{3}}{r} = 3.27 \ \Omega$$

5.3 TRANSFORMER PROTECTION

The power transformer is one of the most important links in a power transmission and distribution system.

- It is a highly reliable piece of equipment. This reliability depends on

- Adequate design

- Careful erection

- Proper maintenance

- Application of protection system.

5.3.1   Protection Equipment Includes

1. Surge diverters

2. Gas relay: It gives early warning of a slowly developing fault, permitting shutdown and repair before severe damage can occur.

3. Electrical relays.

- The choice of suitable protection is also governed by economic considerations. Although this factor is not unique to power transformers, it is brought in prominence by the wide range of transformer ratings used(few KVA to several hundreds MVA)

- Only the simplest protection such as fuses can be justified for transformers of lower ratings.

- For large transformers best protection should be provided.

5.4 TYPES OF FAULTS AFFECTING POWER TRANSFORMER

- Through Faults

  (a) Overload conditions.

  (b) External short-circuit conditions.

- The transformer must be disconnected when such faults occur only after allowing a predetermined time during which other protective gears should have operated.

5.4.1 Internal Faults

- The primary protection of a power transformer is intended for conditions which arises as a result of faults inside the protection zone.

  ➤ Phase-to-earth fault or phase- to- phase fault on HV and LV external terminals

  ➤ Phase-to-earth fault or phase-to- phase fault on HV and LV windings.

  ➤ Interturn faults of HV and LV windings.

  ➤ Earth fault on tertiary winding, or short circuit between turns of a tertiary windings.

  ➤ So called „incipient" faults which are initially minor faults, causing gradually developing fault. These types of faults are not easily detectable at the winding terminals by unbalance current or voltage.

5.4.2 Nature and Effect of Transformer Faults

- A faults on transformer winding is controlled in magnitude by

  ➤ Source and neutral earthling impedance

  ➤ Leakage reactance of the transformer

  ➤ Position of the fault on the winding.

Protection Systems for Transformers

- For protection of generators, Merz-Price circulating-current system is unquestionably the most satisfactory. Though this is largely true of transformer protection, there are cases where circulating current system offers no particular advantage over other systems or impracticable on account of the troublesome conditions imposed by the wide variety of voltages, currents and earthing conditions invariably associated with power transformers. Under such circumstances, alternative protective systems are used which in many cases are as effective as the circulating-current system.

- The principal relays and systems used for transformer protection are :

  - Buchholz devices providing protection against all kinds of incipient faults i.e. slow-developing faults such as insulation failure of windings, core heating, fall of oil level due to leaky joints etc.

  - Earth-fault relays providing protection against earth-faults only.

  - Overcurrent relays providing protection mainly against phase-to-phase faults and overloading.

  - Differential system (or circulating-current system) providing protection against both earth and phase faults.

- The complete protection of transformer usually requires the combination of these systems. Choice of a particular combination of systems may depend upon several factors such as

  - Size of the transformer

  - Type of cooling

  - Location of transformer in the network

  - Nature of load supplied and

  - Importance of service for which transformer is required. In the following sections, above systems of protection will be discussed in detail.

Transformer Differential Protection

Basic discussions related to the Merz-Price Scheme and its limitations which are taken care by the biased differential scheme, are omitted for repetition.

5.5 BUCHHOLZ RELAY

- Buchholz relay is a gas-actuated relay installed in oil immersed transformers for protection against all kinds of faults. Named after its inventor, Buchholz, it is used to give an alarm in case of incipient (i.e. slow-developing) faults in the transformer and to disconnect the transformer from the supply in the event of severe internal faults.

- It is usually installed in the pipe connecting the conservator to the main tank as shown in Fig. 5.15 (a). It is a universal practice to use Buchholz relays on all such oil immersed transformers having ratings in excess of 750 kVA.

Construction :

- Fig. 5.15 (b) shows the constructional details of a Buchholz relay. It takes the form of a domed vessel placed in the connecting pipe between the main tank and the conservator. The device has two elements. The upper element consists of a mercury type switch attached to a float.

- The lower element contains a mercury switch mounted on a hinged type flap located in the direct path of the flow of oil from the transformer to the conservator. The upper element closes an alarm circuit during incipient faults whereas the lower element is arranged to trip the circuit breaker in case of severe internal faults.

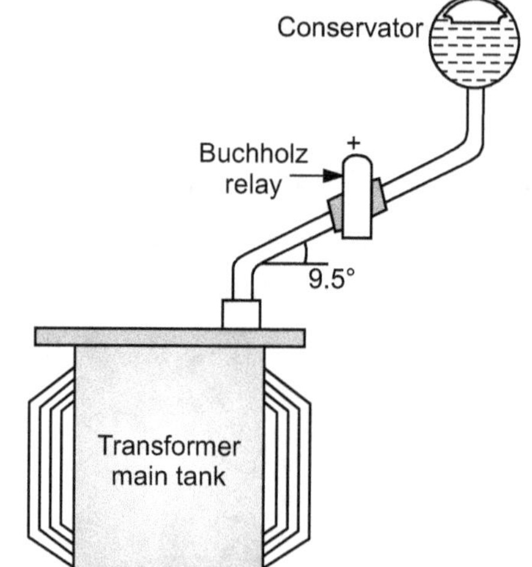

Fig. 5.15 (a)

Fig. 5.15 (b)

Operation :

The operation of Buchholz relay is as follows:

- In case of incipient faults within the transformer, the heat due to fault causes the decomposition of some transformer oil in the main tank. The products of decomposition contain more than 70% of hydrogen gas. The hydrogen gas being light tries to go into the conservator and in the process gets entrapped in the upper part of relay chamber. When a predetermined amount of gas gets accumulated, it exerts sufficient pressure on the float to cause it to tilt and close the contacts of mercury switch attached to it. This completes the alarm circuit to sound an alarm.

- If a serious fault occurs in the transformer, an enormous amount of gas is generated in the main tank. The oil in the main tank rushes towards the conservator via the Buchholz relay and in doing so tilts the flap to close the contacts of mercury switch. This completes the trip circuit to open the circuit breaker controlling the transformer.

Advantages :

- It is the simplest form of transformer protection.
- It detects the incipient faults at a stage much earlier than is possible with other forms of protection.

Disadvantages :

- It can only be used with oil immersed transformers equipped with conservator tanks.
- The device can detect only faults below oil level in the transformer. Therefore, separate protection is needed for connecting cables.

Limitations :

The various limitation of the Buchholz relay are :

- Can be used only for oil immersed transformers having conservator tanks.
- Only faults below oil level are detected.
- Setting of the mercury switches cannot be kept too sensitive otherwise the relay can operate due to bubbles, vibration, earthquakes mechanical shocks etc.
- The relay is slow to operate having minimum operating time of 0.1 seconds and average time of 0.2 seconds.

Applications

- The following types of transformer faults can be protected by the Buchholz relay and are indicated by alarm:

  1. Local overheating
  2. Entrance of air bubbles in oil
  3. Core bolt insulation failure
  4. Short circuited laminations
  5. Loss of oil and reduction in oil level due to leakage
  6. Bad and loose electrical contacts
  7. Short circuit between phases
  8. Winding short circuit
  9. Bushing puncture
  10. Winding earth fault.

5.6 EARTH-FAULT PROTECTION

5.6.1 Earth-Fault or Leakage Protection

- An earth-fault usually involves a partial breakdown of winding insulation to earth. The resulting leakage current is considerably less than the short-circuit current. The earth-fault may continue for a long time and cause considerable damage before it ultimately develops into a short-circuit and removed from the system.

- Under these circumstances, it is profitable to employ earth-fault relays in order to ensure the disconnection of earth-fault or leak in the early stage. An earth-fault relay is essentially an overcurrent relay of low setting and operates as soon as an earth-fault or leak develops. One method of protection against earth-faults in a transformer is the core balance leakage protection shown in Fig. 5.16.

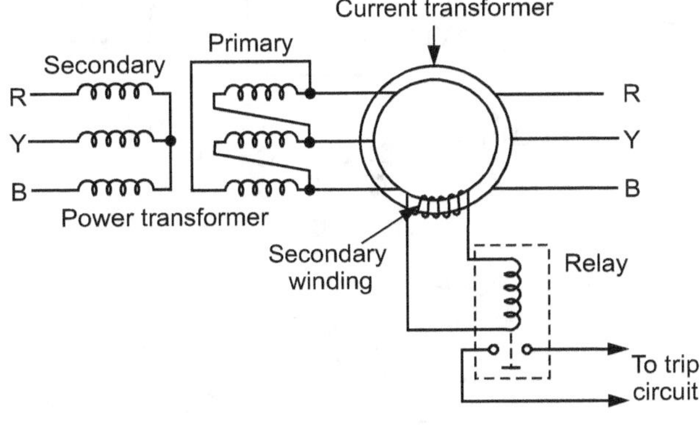

Fig. 5.16

- The three leads of the primary winding of power transformer are taken through the core of a current transformer which carries a single secondary winding. The operating coil of a relay is connected to this secondary. Under normal conditions (i.e. no fault to earth), the vector sum of the three phase currents is zero and there is no resultant flux in the core of current transformer no matter how much the load is out of balance.

- Consequently, no current flows through the relay and it remains inoperative. However, on the occurrence of an earth-fault, the vector sum of three phase currents is no longer zero. The resultant current sets up flux in the core of the C.T. which induces e.m.f. in the secondary winding. This energises the relay to trip the circuit breaker and disconnect the faulty transformer from the system.

5.6.2 Combined Leakage and Overload Protection

- The core-balance protection described above suffers from the drawback that it cannot provide protection against overloads. If a fault or leakage occurs between phases, the core-balance relay will not operate. It is a usual practice to provide combined leakage and overload protection for transformers. The earth relay has low current setting and operates under earth or leakage faults only. The overload relays have high current setting and are arranged to operate against faults between the phases.

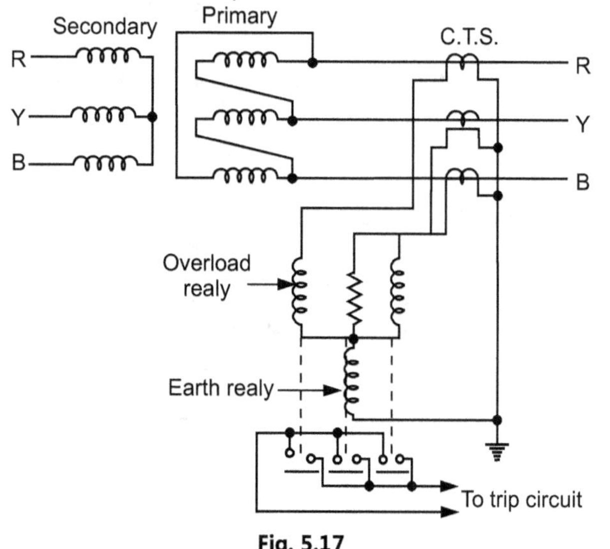

Fig. 5.17

- Fig. 5.17 shows the schematic arrangement of combined leakage and overload protection. In this system of protection, two overload relays and one leakage or earth relay are connected as shown. The two overload relays are sufficient to protect against phase-to-phase faults. The trip contacts of overload relays and earthfault relay are connected in parallel. Therefore, with the energising of either overload relay or earth relay, the circuit breaker will be tripped.

5.7 TRANSFORMER DIFFERENTIAL PROTECTION

- Basic discussions related to the Merz-Price Scheme and its limitations which are taken care by the biased differential scheme, are omitted for repetition.

5.7.1 Basic Considerations

1. Transformation Ratio

- The nominal currents in the primary and secondary sides of the transformer vary in inverse ratio to the corresponding voltages. This should be compensated for by using different transformation ratios for the CTs on the primary and secondary sides of the transformer.

2. Current Transformer Connections

- When a transformer is connected in star/delta, the secondary current has a phase shift of 300 relative to the primary

- This phase shift can be offset by suitable secondary CT connections

- The zero-sequence currents flowing on the star-side of the transformer will not produce current outside the delta on the other side. The zero sequence current must therefore be eliminated from the star-side by connecting the CTs in delta.

- The CTs on delta side should be connected in star in order to give 300 phase shift.

- When CTs are connected in delta, their secondary ratings must be reduced to 1/3 times the secondary ratings of the star-connected transformer, in order that the currents outside the delta may balance with the secondary currents of the star-connected CTs.

- If transformers were connected in star/star, the CTs on both sides would need be connected in delta-delta.

3. Bias to cover tap-changing facility and CT mismatch

- If the transformer has the benefit of a tap changer, it is possible to vary its transformation ratio for voltage control.

- The differential protection system should be able to cope with this variation.

- This is because if the CTs are chosen to balance for the mean ratio of the power transformer, a variation in ratio from the mean will create an unbalance proportional to the ratio change. At maximum through fault current, the spill output produced by the small percentage unbalance may be substantial

- Differential protection should be provided with a proportional bias of an amount which exceeds in effect the maximum ratio deviation. This stabilizes the protection under through fault conditions while still permitting the system to have good basic sensitivity

4. Magnetization Inrush

- The magnetizing inrush produces a current flow into the primary winding that does not have any equivalent in the secondary winding. The net effect is thus similar to the situation when there is an internal fault on the transformer.

- Since the differential relay sees the magnetizing current as an internal fault, it is necessary to have some method of distinguishing between the magnetizing current and the fault current using one or all of the following methods.

- Using a differential relay with a suitable sensitivity to cope with the magnetizing current, usually obtained by a unit that introduces a time delay to cover the period of the initial inrush peak.

- Using a harmonic-restraint unit, or a supervisory unit, in conjunction with a differential unit.

- Inhibiting the differential relay during the energizing the transformer.

- Compared to the differential protection used in generators, there are certain important points discussed below which must be taken care of while using such protection for the power transformers :

1. In a power transformer, the voltage rating of the two windings is different. The high voltage winding is low current winding while low voltage winding is high current winding. Thus there always exists difference in current on the primary and secondary sides of the power transformer. Hence if C.T.s of same ratio are used on two sides, then relay may get operated through there is no fault existing.

- To compensate for this difficulty, the current ratios of C.T.s on each side are different. These ratios depend on the line currents of the power transformer and the connection of C.T.s. Due to the different turns ratio, the currents fed into the pilot wires from each end are same under normal conditions so that the relay remains inoperative. For example if K is the turns ratio of a power transformer then the ratio of C.T.s on low voltage side is made K times greater than that of C.T.s on high voltage side.

2. In case of power transformers, there is an inherent phase difference between the voltages induced in high voltage winding and low voltage winding. Due to this, there exists a phase difference between the line currents on primary and secondary sides of a power transformer. This introduces the phase difference between the C.T. secondary currents, on the two sides of a power transformer. Through the turns ratio of C.T.s are selected to compensate for turns ratio of transformer, a differential current may result due to the phase difference between the currents on two sides. Such a different current may operate the relay though there is no fault. Hence it is necessary to correct the phase difference.

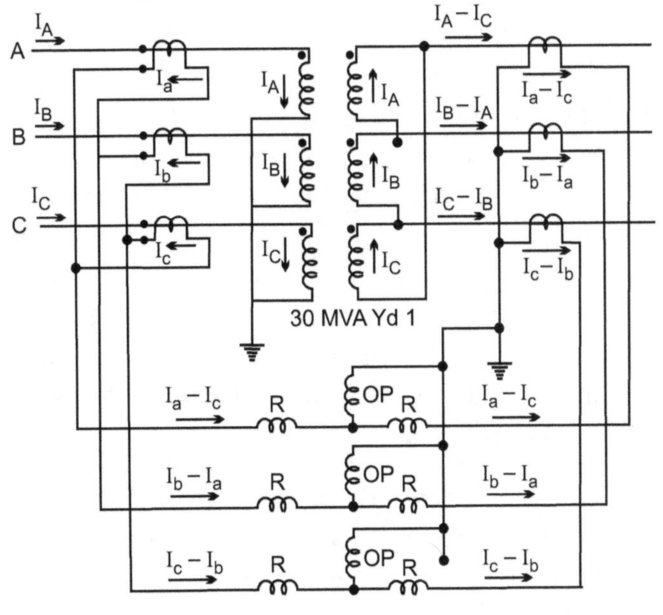

Fig. 5.18

- To compensate for this, the C.T. connections should be such that the resultant currents fed into the pilot wires from either sides are displaced in phase by an angle equal to the phase shift between the primary and secondary currents. To achieve this, secondaries of C.T.s on star connected side of a power transformer are connected in delta while the secondaries of C.T.s on delta connected side of a power transformer are connected in star.

- The table gives the way of connecting C.T. secondaries for the various types of power transformer connections.

Power Transformer Connections		C. T. Connections	
Primary	Secondary	Primary	Secondary
Star	Delta	Delta	Star
Delta	Delta	Star	Star
Star	Star	Delta	Delta
Delta	Start	Star	Delta

- With such an arrangement, the phase displacement between the currents gets compensated with the oppositely connected C.T. secondaries. Hence currents fed to the pilot wires from both the sides are in phase under normal running conditions and the relay is ensured to be inoperative.

3. The neutrals of C.T. star and power transformer stars are grounded.

4. Many transformers have tap changing arrangement due to which there is a possibility of flow of differential current. For this, the turns ratio of C.T.s on both sides of the power transformer are provided with tap for of C.T.s on both sides of the power transformer are provided with tap for their adjustment. For the sake of understanding, the connection of C.T. secondaries in delta for star side of power transformer and the connection of C.T. secondaries in star for delta

5.7.2 Star/Deltaunit

- Let us study the Differential protection for the star-delta power transformer. The primary of the power transformer is star connected while the secondary is delta connected. Hence to compensate for the phase difference, the C.T. secondaries on primary side must

be connected in delta while the C.T. secondaries on delta side must be connected in star.

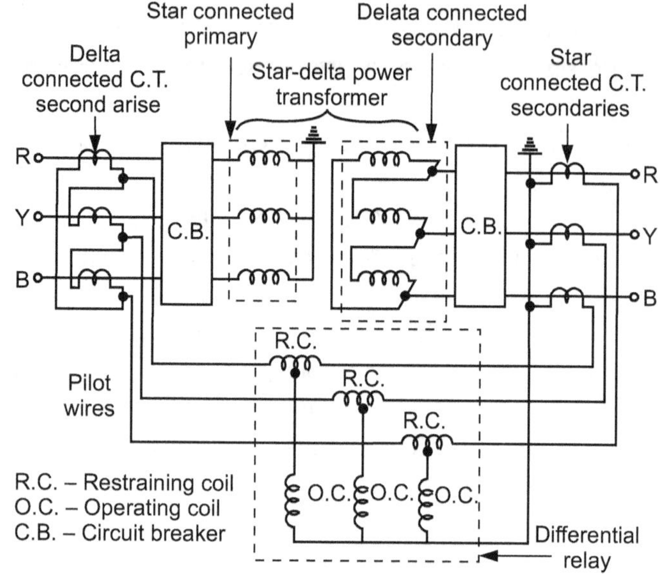

Fig. 5.19

- The star point of the power transformer primary as well as the star connected C.T. secondaries must be grounded. The restraining coils are connected across the C.T. secondary windings while the operating coils are connected between the tapping points on the restraining coils and the star point of C.T. secondaries.

- With the proper selection of turns ratio of C.T.s the coils are under balanced condition during normal operating conditions. The C.T. secondaries carry equal currents which are in phase under normal conditions. So no current flows through the relay and the relay is inoperative.

- It is important to note that this scheme gives protection against short circuit faults between the turns i.e. interturn faults also. This is because when there is an interturn fault, the turns ratio of power transformer gets affected. Due to this the currents on both sides of the power transformer become unbalanced. This causes an enough differential current which floes through the relay and the relay operates.

5.7.3 Star/Starunit

- The Fig. 5.20 above shows the Merz-Price protection system for the star-star power transformer. Both primary and secondary of the power transformer are connected in star and hence C.T. secondaries.

- The operating coils are connected between the tapping on the restraining coil and the ground. The operation of the scheme remains same for any type of power transformer as discussed for star-star power transformer.

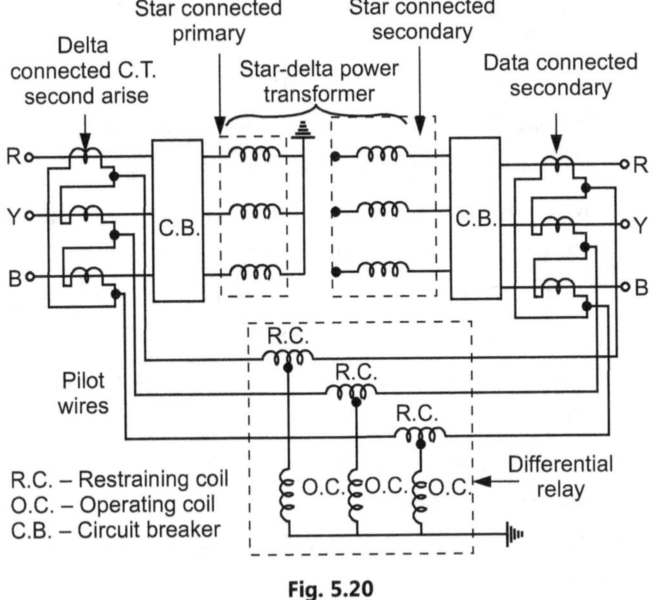

Fig. 5.20

Example 5.6 : *A 3-phase transformer of 220/11,000 line volts is connected in star/delta. The protective transformer on 220 V side have a current ratio of 600/5. What should be the CT ratio on 11000 V side ?*

Solution : For star / delta power transformers, CTs will be connected in delta on 220 V side (i.e. star side of power transformer) and in star an 11,000 V side (i.e. delta side of power transformer) as shown in Fig. 5.21.

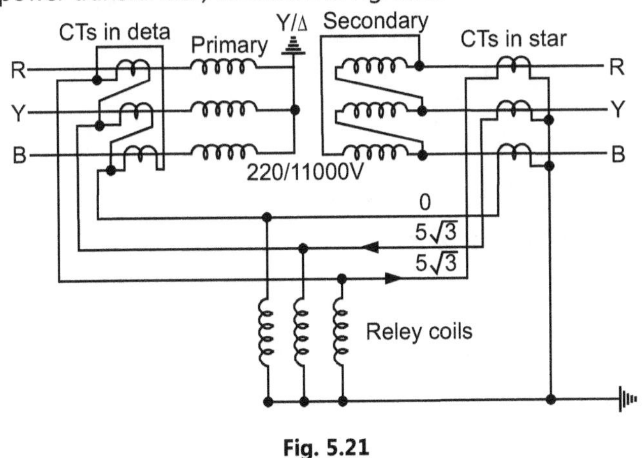

Fig. 5.21

Suppose that line current on 220 V side is 600 A.

∴ Phase current of delta connected CTs on 220 V side

$$= 5A$$

Line current of delta connected CTs on 220 V side

$$= 5 \times \sqrt{3} = 5\sqrt{3} \text{ A}$$

This current (i.e. $5\sqrt{3}$) will flow through the pilot wires. Obviously, this will be the current which flows through the secondary of CTs on the 11,000 V side.

∴ Phase current of star connected CTs on 11,000 V side

$$= 5\sqrt{3} \text{ A}.$$

If I is the line current on 11,000 V side, then

Primary apparent power = Secondary apparent power

or $\sqrt{3} \times 220 \times 600 = \sqrt{3} \times 11,000 \times I$

or $I = \dfrac{\sqrt{3} \times 220 \times 600}{\sqrt{3} \times 11000} = 12 \text{ A}$

∴ Turn-ratio of CTs on 11000 V side

$$= 12 : 5\sqrt{3} = 1.385 : 1$$

Example 5.7 : *A 3-phase transformer having line-voltage ratio of 0.4 kV/11 kV is connected in star-delta and protective transformers on the 400 V side have a current ratio of 500/5. What must be the ratio of the protective transformers on the 11 kV side?*

Solution : Fig. 5.22 shows the circuit connections. For star/delta transformers, CTs will be connected in delta on 400 V side (i.e. star side of power transformer) and in star on 11000 V side (i.e. delta side of power transformer).

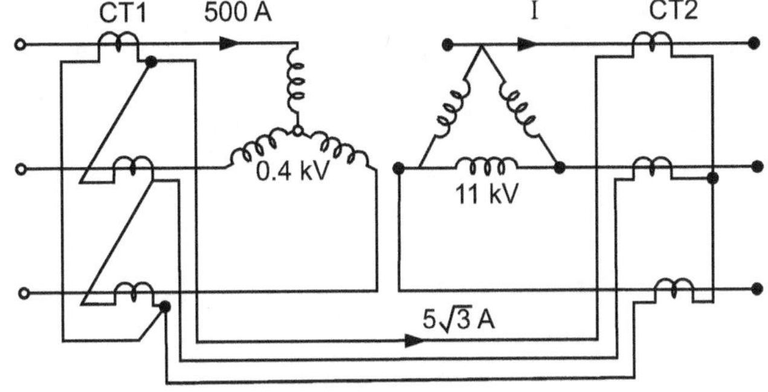

Fig. 5.22

Suppose the line current on 400 V side is 500 A.

∴ Phase current of delta connected CTs on 400 V side

$$= 5A$$

Line current of delta connected CTs on 400 V side

$$= 5 \times \sqrt{3} = 5\sqrt{3}A$$

This current (i.e. $5\sqrt{3}A$) will flow through the pilot wires. Obviously, this will be the current which flows through the secondary of the CTs on 11000 V side.

∴ Phase current of star-connected CTs on 11000 V side.

$$= 5\sqrt{3}A$$

If I is the line current on 11000 V side, then,

Primary apparent power = Secondary apparent power

or $\qquad \sqrt{3} \times 400 \times 500 = \sqrt{3} \times 11000 \times I$

or $\qquad I = \dfrac{\sqrt{3} \times 400 \times 500}{\sqrt{3} \times 11000} = \dfrac{200}{11}$ A

∴   C.R. ratio of CTs on 11000 V side

$$= \dfrac{200}{11} : 5\sqrt{3} = \dfrac{200}{11 \times 5\sqrt{3}}$$

$$= \dfrac{10.5}{5} = 10.5 : 5$$

EXERCISE

1. Discuss the important faults on an Alternator.

2. Explain with a neat diagram the application of Merz-Price circulating current principle for the protection of alternator.

3. Describe with a neat diagram the balanced earth protection for small-size generators.

4. Describe the construction and working of a Buchholz relay.

5. Write short notes on the following :

   (i) Earth-fault protection for alternator

   (ii) Combined leakage and overload protection for transformers

   (iii) Earth-fault protection for transformers

6. How will you protect an alternator from turn-to-turn fault on the same phase winding ?

7. What factors cause difficulty in applying circulating current principle to a power transformer ?

8. Describe the Merz-Price circulating current system for the protection of transformers.

9. Why is overload protection not necessary for alternators ?

10. What is the difference between an earth relay and overcurrent relay ?

11. How many faults develop in a power transformer ?

INSULATION CO-ORDINATION

6.1 INTRODUCTION

- Insulation coordination means the correlation of the insulation of the various equipments in a power system to the insulation of the protective devices used for the protection of those equipments against overvoltages.

- In a power system various equipments like transformers, circuit breakers, bus supports etc. have different breakdown voltages and hence the volt-time characteristics. In order that all the equipments should be properly protected it is desired that the insulation of the various protective devices must be properly coordinated.

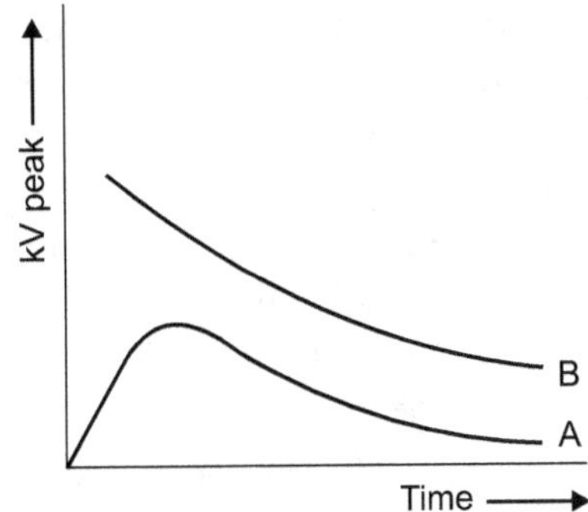

Fig. 6.1: Volt-time curve A (protecting device and) volt-time curve B (device to be protected)

- The basic concept of insulation coordination is illustrated in Fig. 6.1. Curve A is the volt-time Curve of the protective device and B the volt-time curve of the equipment to be protected. Figure 6.1 shows the desired positions of the volt-time curves of the protecting device and the equipment to be protected. Thus, any insulation having a withstand voltage strength in excess of the insulation strength of curve B is protected by the protective device of curve A.

- The 'volt-time curve' expression will be used very frequently in this chapter. It is, therefore, necessary to understand the meaning of this expression.

6.1.1 Volt-Time Curve

- The breakdown voltage for a particular insulation or flashover voltage for a gap is a function of both the magnitude of voltage and the time of application of the voltage. The volt-time curve is a graph showing the relation between the crest flashover voltages and the time to flashover for a series of impulse applications of a given wave shape. For the construction of volt-time curve the following procedure is adopted. Waves of the same shape but of different peak values are applied to the insulation whose volt-time curve is required.

- If flashover occurs on the front of the wave, the flashover point gives one point on the volt-time curve. The other possibility is that the flashover occurs just at the peak value of the wave; this gives another point on the V-T curve. The third possibility is that the flashover occurs on the tail side of the wave.

- In this case to find the point on the V-T curve, draw a horizontal line from the peak value of this wave and also draw a vertical line passing through the point where the flashover takes place. The intersection of the horizontal and vertical lines gives the point on the V-T curve. This procedure is nicely shown in Fig. 6.2.

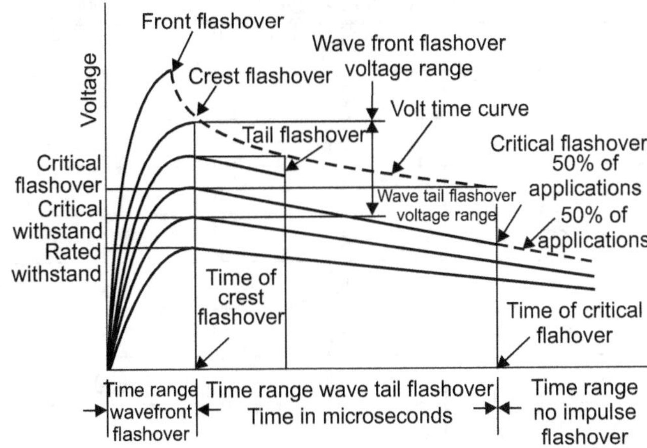

Fig. 6.2 : Volt-time curve (construction)

- The overvoltages against which coordination is required could be caused on the system due to system faults, switching operation or lightning surges. For lower voltages, normally up to about 345 kV, overvoltages caused by system faults or switching operations do not cause damage to equipment insulation although they may be detrimental to protective devices. Overvoltages caused by lightning are of sufficient magnitude to affect the equipment insulation whereas for voltages above 345 kV it is these switching surges which are more dangerous for the equipments than the lightning surges.

- The problem of coordinating the insulation of the protective equipment involves not only guarding the equipment insulation but also it is desired that the protecting equipment should not be damaged.

To assist in the process of insulation coordination, standard insulation levels have been recommended. These insulation levels are defined as follows:

- Basic impulse insulation levels (BIL) are reference levels expressed in impulse crest voltage with a standard wave not longer than 1.2/50 µsec wave. Apparatus insulation as demonstrated by suitable tests shall be equal to or greater than the basic insulation level.

6.1.2 The Problem of Insulation Co-ordination can be Studied Under Three Steps

1. Selection of a suitable insulation which is a function of reference class voltage (i.e., 1.05 × operating voltage of the system). Table 6.1 gives the BIL for various reference class voltages.

Table 6.1 : Basic Impulse Insulation Levels

Reference Class kV	Standard Basic Impulse Level kV	Reduced Insulation Levels
23	150	
34.5	200	
46	250	
69	350	
92	450	
115	550	450
138	650	550
161	750	650
196	900	
230	1050	900
287	1300	1050
345	1550	1300

2. The design of the various equipments such that the breakdown or flashover strength of all insulation in the station equals or exceeds the selected level as in 6.1.

3. Selection of protective devices that will give the apparatus as good protection as can be justified economically.

- The above procedure requires that the apparatus to be protected shall have a withstand test value not less than the kV magnitude given in the second column of Table 6.1, irrespective of the polarity of the wave positive or negative and irrespective of how the system was grounded.

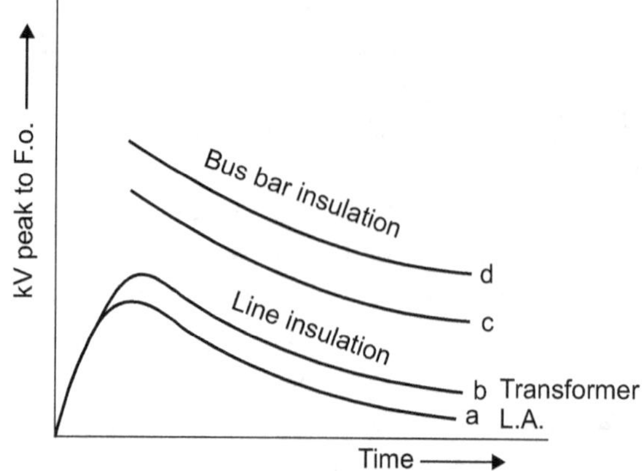

Fig. 6.3

- The third column of the table gives the reduced insulation levels which are used for selecting insulation levels of solidly grounded systems and for systems operating above 345 kV where switching surges are of more importance than the lightning surges. At 345 kV, the switching voltage is considered to be 2.7 p.u., i.e., 345 × 2.7 = 931.5 kV which corresponds to the lightning level. At 500 kV, however, 2.7 p.u. will mean 2.7 × 500 = 1350 kV switching voltage which exceeds the lightning voltage level.

- Therefore, the ratio of switching voltage to operating voltage is reduced by using the switching resistances between the C.B. contacts. For 500 kV, it has been possible to obtain this ratio as 2.0 and for 765 kV it is 1.7. With further increase in operating voltages it is hoped that the ratio could be brought to 1.5. So, for switching voltages the reduced levels in third column are used i.e., for 345 kV, the standard BIL is 1550 kV but if the equipment can withstand even 1425 kV or 1300 kV it will serve the purpose.

- Fig. 6.3 gives the relative position of the volt-time curves of the various equipments in a substation for proper coordination. To illustrate the selection of the BIL of a transformer to be operated on a 138 kV system assume that the transformer is of large capacity and its star point is solidly grounded. The grounding is such that the line-to-ground voltage of the healthy phase during a ground fault on one of the phase is say 74% of the normal L-L voltage.

- Allowing for 5% overvoltage during operating conditions, the arrester rms operating voltage will be 1.05 × 0.74 × 138 = 107.2 kV. The nearest standard rating is 109 kV. The characteristic of such a L.A. is shown in Fig. 6.4. From the figure the breakdown value of the arrester is 400 kV. Assuming a 15% margin plus 35 kV between the insulation levels of L.A. and the transformer, the insulation level of transformer should be at least equal to 400 + 0.15 × 400 + 35 = 495 kV. From Fig. 6.4 (or from the table the reduced level of transformer for 138 kV is 550 kV) the insulation level of transformer is 550 kV; therefore a lightning arrester of 109 kV rating can be applied.

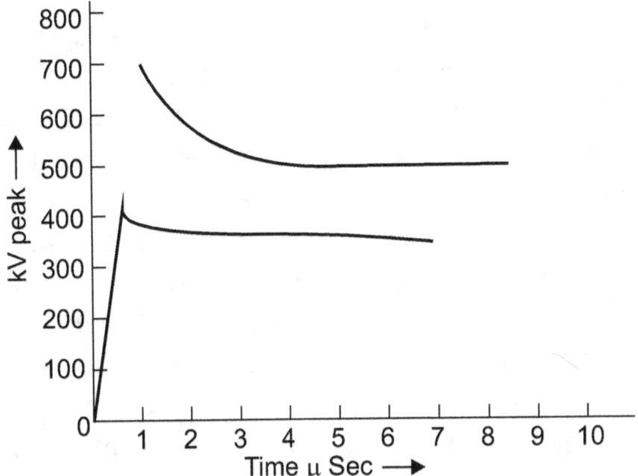

Fig. 6.4 : Coordination of transformer insulation with lightning arrester : A – Lightning arrester 109 kV, B – transformer insulation withstand characteristic

- It is to be noted that low voltage lines are not as highly insulated as higher voltage lines so that lightning surges coming into the station would normally be much less than in a higher voltage station because the high voltage surges will flashover the line insulation of low voltage line and not reach the station.

- The traditional approach to insulation coordination requires the evaluation of the highest overvoltages to which an equipment may be subjected during operation and selection of standardized value of withstand impulse voltage with suitable safety margin. However, it is realized that overvoltages are a random phenomenon and it is uneconomical to design plant with such a high degree of safety that they sustain the infrequent ones.

- It is also known that insulation designed on this basis does not give 100% protection and insulation failure may occur even in well designed plants and, therefore, it is desired to limit the frequency of insulation failures to the most economical value taking into account equipment cost and service continuity. Insulation coordination, therefore, should be based on evaluation and limitation of the risk of failure than on the a priori choice of a safety margin.

- The modern practice, therefore, is to make use of probabilistic concepts and statistical procedures especially for very high voltage equipments which might later on be extended to all cases where a close adjustment of insulation to system conditions proves economical. The statistical methods even though laborious are quite useful.

6.2 FLASHOVER VOLTAGE

The voltage at which an electric discharge occurs between two electrodes that are separated by an insulator; the value depends on whether the insulator surface is dry or wet. Also known as sparkover voltage.

6.2.1 Wet Flashover Voltage

- The voltage at which an electric discharge occurs between two electrodes that are separated by an insulator whose surface has been sprayed with water to simulate rain.

6.2.2 Dry Flashover Voltage

- Voltage at which the air surrounding a clean dry insulator or shell completely breaks down between electrodes.

6.2.3 Impulse Flashover Voltage

- Impulse voltage are very short high voltage or current that are used to test the strength of electric power equipment against lightning and switching surges.

6.3 CAUSES OF OVERVOLTAGES

The overvoltages on a power system may be broadly divided into two main categories viz.

1. **Internal Causes**
   (i)   Switching surges
   (ii) Insulation failure
   (iii) Arcing ground
   (iv) Resonance

2. **External Causes i.e. Lightning**

- Internal causes do not produce surges of large magnitude. Experience shows that surges due to internal causes hardly increase the system voltage to twice the normal value. Generally, surges due to internal causes are taken care of by providing proper insulation to the equipment in the power system.

- However, surges due to lightning are very severe and may increase the system voltage to several times the normal value. If the equipment in the power system is not protected against lightning surges, these surges may cause considerable damage. In fact, in a power system, the protective devices provided against overvoltages mainly take care of lightning surges

6.3.1 Internal Causes of Overvoltages

- Internal causes of overvoltages on the power system are primarily due to oscillations set up by the sudden changes in the circuit conditions. This circuit change may be a normal switching operation such as opening of a circuit breaker, or it may be the fault condition such as grounding of a line conductor. In practice, the normal system insulation is suitably designed to withstand such surges. We shall briefly discuss the internal causes of overvoltages.

1. **Switching Surges :** The overvoltages produced on the power system due to switching operations are known as switching surges. A few cases will be discussed by way of illustration :

(i) **Case of an Open Line :** During switching operations of an unloaded line, travelling waves are set up which produce overvoltages on the line. As an illustration, consider an unloaded line being connected to a voltage source as shown in Fig. 6.5.

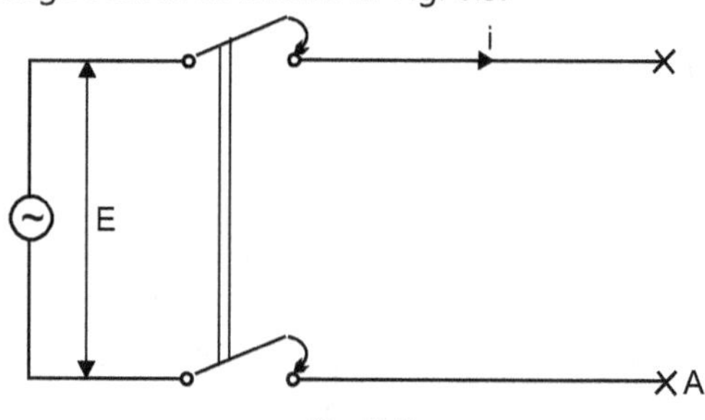

Fig. 6.5

- When the unloaded line is connected to the voltage source, a voltage wave is set up which travels along the line. On reaching the terminal point A, it is reflected back to the supply end without change of sign. This causes voltage doubling i.e. voltage on the line becomes twice the normal value. If E r.m.s. is the supply voltage, then instantaneous voltage which the line will have to withstand will be $2\sqrt{2}\,E$.

- This overvoltage is of temporary nature. It is because the line losses attenuate the wave and in a very short time, the line settles down to its normal supply voltage E. Similarly, if an unloaded line is switched off, the line will attain a voltage of $2\sqrt{2}\,E$ for a moment before settling down to the normal value.

(ii) **Case of a Loaded Line :** Overvoltages will also be produced during the switching operations of a loaded line. Suppose a loaded line is suddenly interrupted. This will set up a voltage of $2\,Z_n\,i$ across the break (i.e. switch) where i is the instantaneous value of current at the time of opening of line and Z_n is 1000 Ω carries a

current of 100 A (r.m.s) and the break occurs at the moment when current is maximum. The voltage across the breaker (i.e. switch) $= 2\sqrt{2} \times 100 \times \dfrac{1000}{1000} = 282.8$ kV. If V_m is the peak value of voltage in kV, the maximum voltage to which the line may be subjected is $= (V_m^m + 282.8)$ kV.

(iii) **Current Chopping :** Current chopping results in the production of high voltage transients across the contacts of the air blast circuit breaker. It is briefly discussed here. Unlike oil circuit breakers, which are independent for the effectiveness on the magnitude of the current being interrupted, air-blast circuit breakers retain the same extinguishing power irrespective of the magnitude of this current. When breaking low currents (e.g. transformer magnetising current) with air-blast breaker, the powerful de-ionising effect of air-blast causes the current to fall abruptly to zero well before the natural current zero is reached. This phenomenon is called current chopping and produces high transient voltage across the breaker contacts. Overvoltages due to current chopping are prevented by resistance switching

2. **Insulation Failure :** The most common case of insulation failure in a power system is the grounding of conductor (i.e. insulation failure between line and earth) which may cause overvoltages in the system. This is illustrated in Fig. 6.6.

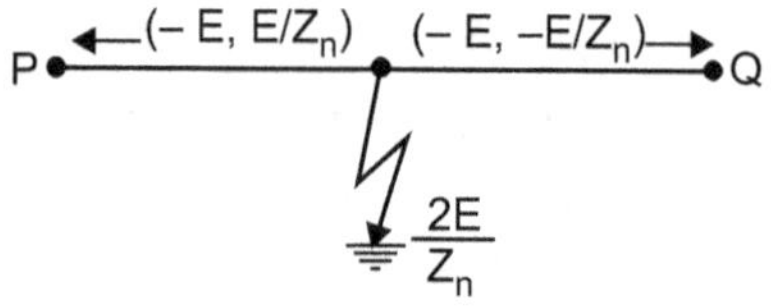

Fig. 6.6

Suppose a line at potential E is earthed at point X. The earthing of the line causes two equal voltages of –E to travel along XQ and XP containing currents –E/Zn and +E/Zn respectively. Both these currents pass through X to earth so that current to earth is 2 E/Zn.

3. **Arcing Ground :** In the early days of transmission, the neutral of three phase lines was not earthed to gain two advantages. Firstly, in case of line-to-ground fault, the line is not put out of action. Secondly, the zero sequence currents are eliminated, resulting in the decrease of interference with communication lines. Insulated neutrals give no problem with short lines and comparatively low voltages. However, when the lines are long and operate at high voltages, serious problem called arcing ground is often witnessed. The arcing

ground produces severe oscillations of three to four times the normal voltage.

The phenomenon of intermittent arc taking place in line-to-ground fault of a 3φ system with consequent production of transients is known as arcing ground. The transients produced due to arcing ground are cumulative and may cause serious damage to the equipment in the power system by causing breakdown of insulation. Arcing ground can be prevented by earthing the neutral.

4. **Resonance :** Resonance in an electrical system occurs when inductive reactance of the circuit becomes equal to capacitive reactance. Under resonance, the impedance of the circuit is equal to resistance of the circuit and the p.f. is unity. Resonance causes high voltages in the electrical system. In the usual transmission lines, the capacitance is very small so that resonance rarely occurs at the fundamental supply frequency. However, if generator e.m.f. wave is distorted, the trouble of resonance may occur due to 5th or higher harmonics and in case of underground cables too.

6.3.2 External Causes i.e. Lightning

- An electric discharge between cloud and earth, between clouds or between the charge centres of the same cloud is known as lightning.

- Lightning is a huge spark and takes place when clouds are charged to such a high potential (+ve or −ve) with respect to earth or a neighbouring cloud that the dielectric strength of neighbouring medium (air) is destroyed. There are several theories which exist to explain how the clouds acquire charge. The most accepted one is that during the uprush of warm moist air from earth, the friction between the air and the tiny particles of water causes the building up of charges.

- When drops of water are formed, the larger drops become positively charged and the smaller drops become negatively charged. When the drops of water accumulate, they form clouds, and hence cloud may possess either a positive or a negative charge, depending upon the charge of drops of water they contain. The charge on a cloud may become so great that it may discharge to another cloud or to earth and we call this discharge as lightning. The thunder which accompanies lightning is due to the fact that lightning suddenly heats up the air, thereby causing it to expand. The surrounding air pushes the expanded air back and forth causing the wave motion of air which we recognise as thunder

Mechanism of Lightning Discharge

- Let us now discuss the manner in which a lightning discharge occurs. When a charged cloud passes over the earth, it induces equal and opposite charge on the earth below. Fig.6.7 shows a negatively charged cloud inducing a positive charge on the earth below it.

- As the charge acquired by the cloud increases, the potential between cloud and earth increases and, therefore, gradient in the air increases. When the potential gradient is sufficient (5 kV/cm to 10 kV/cm) to break down the surrounding air, the lightning stroke starts. The stroke mechanism is as under :

(i) As soon as the air near the cloud breaks down, a streamer called leader streamer or pilot streamer starts from the cloud towards the earth and carries charge with it as shown in Fig. 6.7 (a). The leader streamer will continue its journey towards earth as long as the cloud, from which it originates feeds enough charge to it to maintain gradient at the tip of leader streamer above the strength of air. If this gradient is not maintained, the leader streamer stops and the charge is dissipated without the formation of a complete stroke. In other words, the leader streamer will not reach the earth. Fig. 6.7 (a) shows the leader streamer being unable to reach the earth as gradient at its end cloud not be maintained above the strength of air. It may be noted that current in the leader streamer is low (<100 A) and its velocity of propagation is about 0·05% that of velocity of light. Moreover, the luminosity of leader is also very low.

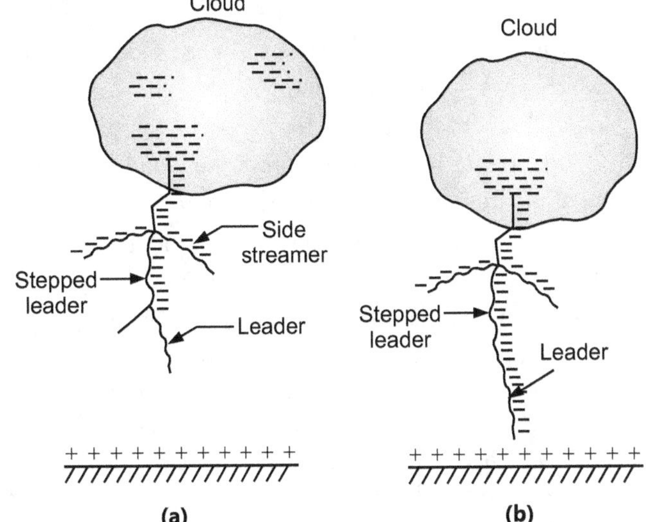

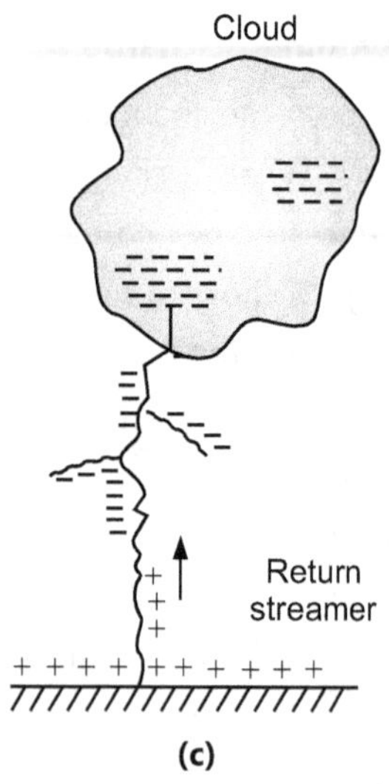

(c)

Fig. 6.7

(ii) In many cases, the leader streamer continues its journey towards earth [See Fig. 6.7 (b)] until it makes contact with earth or some object on the earth. As the leader streamer moves towards earth, it is accompanied by points of luminescence which travel in jumps giving rise to stepped leaders. The velocity of stepped leader exceeds one-sixth of that of light and distance travelled in one step is about 50 m. It may be noted that stepped leaders have sufficient luminosity and give rise to first visual phenomenon of discharge.

(iii) The path of leader streamer is a path of ionisation and, therefore, of complete breakdown of insulation. As the leader streamer reaches near the earth, a return streamer shoots up from the earth [See Fig. 6.7 (c)] to the cloud, following the same path as the main channel of the downward leader. The action can be compared with the closing of a switch between the positive and negative terminals; the downward leader having negative charge and return streamer the positive charge. This phenomenon causes a sudden spark which we call lightning. With the resulting neutralisation of much of the negative charge on the cloud, any further discharge from the cloud may have to originate from some other portion of it.

The following points may be noted about lightning discharge :

- A lightning discharge which usually appears to the eye as a single flash is in reality made up of a number of separate strokes that travel down the same path. The interval between them varies from 0·0005 to 0·5 second. Each separate stroke starts as a downward leader from the cloud.

- It has been found that 87% of all lightning strokes result from negatively charged clouds and only 13% originate from positively charged clouds.

- It has been estimated that throughout the world, there occur about 100 lightning strokes per second.

- Lightning discharge may have currents in the range of 10 kA to 90 kA.

6.4 LIGHTNING STROKES

6.4.1 Types of Lightning Strokes

There are two main ways in which a lightning may strike the power system (e.g. overhead lines, towers, sub-stations etc.), namely;

1. Direct stroke
2. Indirect stroke

1. **Direct Stroke :** In the direct stroke, the lightning discharge (i.e. current path) is directly from the cloud to the subject equipment e.g. an overhead line. From the line, the current path may be over the insulators down the pole to the ground. The overvoltages set up due to the stroke may be large enough to flashover this path directly to the ground. The direct strokes can be of two types viz. (i) Stroke A and (ii) stroke B.

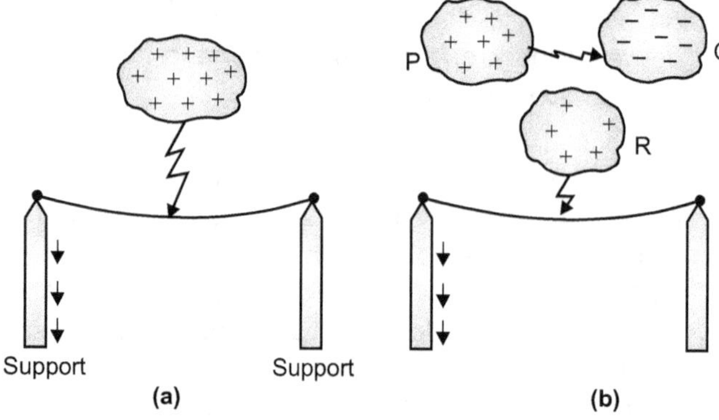

Fig. 6.8

- In stroke A, the lightning discharge is from the cloud to the subject equipment i.e. an overhead line in this case as shown in Fig. 6.8 (a). The cloud will induce a charge of opposite sign on the tall object (e.g. an overhead line in this case). When the potential between the cloud and line exceeds the breakdown value of air, the lightning discharge occurs between the cloud and the line.

- In stroke B, the lightning discharge occurs on the overhead line as a result of stroke A between the clouds as shown in Fig. 6.8 (b). There are three clouds P, Q and R having positive, negative and positive charges respectively. The charge on the cloud Q is bound by the cloud R. If the cloud P shifts too near the cloud Q, then lightning discharge will occur between them and charges on both these clouds disappear quickly. The result is that charge on cloud R suddenly becomes free and it then discharges rapidly to earth, ignoring tall objects.

Two points are worth noting about direct strokes. Firstly, direct strokes on the power system are very rare. Secondly, stroke A will always occur on tall objects and hence protection can be provided against it. However, stroke B completely ignores the height of the object and can even strike the ground. Therefore, it is not possible to provide protection against stroke B.

2. **Indirect Stroke :** Indirect strokes result from the electrostatically induced charges on the conductors due to the presence of charged clouds. This is illustrated in Fig. 6.9. A positively charged cloud is above the line and induces a negative charge on the line by electrostatic induction. This negative charge, however, will be only on that portion of the line right under the cloud and the portions of the line away from it will be positively charged as shown in Fig. 6.9. The induced positive charge leaks slowly to earth via the insulators. When the cloud discharges to earth or to another cloud, the negative charge on the wire is isolated as it cannot flow quickly to earth over the insulators. The result is that negative charge rushes along the line is both directions in the form of travelling waves. It may be worthwhile to mention here that majority of the surges in a transmission line are caused by indirect lightning strokes.

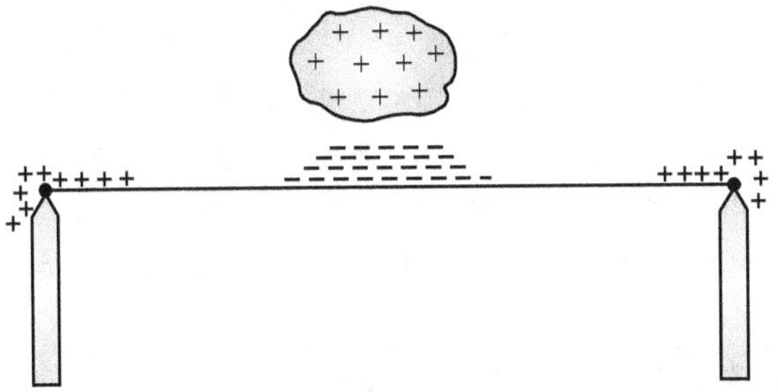

Fig. 6.9

6.4.2 Harmful Effects of Lightning

A direct or indirect lightning stroke on a transmission line produces a steep-fronted voltage wave on the line. The voltage of this wave may rise from zero to peak value (perhaps 2000 kV) in about 1 µs and decay to half the peak value in about 5µs. Such a steep-fronted voltage wave will initiate travelling waves along the line in both directions with the velocity dependent upon the L and C parameters of the line.

- The travelling waves produced due to lightning surges will shatter the insulators and may even wreck poles.

- If the travelling waves produced due to lightning hit the windings of a transformer or generator, it may cause considerable damage. The inductance of the windings opposes any sudden passage of electric charge through it. Therefore, the electric charges "piles up" against the transformer (or generator). This induces such an excessive pressure between the windings that insulation may breakdown, resulting in the production of arc. While the normal voltage between the turns is never enough to start an arc, once the insulation has broken down and an arc has been started by a momentary overvoltage, the line voltage is usually sufficient to maintain the arc long enough to severely damage the machine.

- If the arc is initiated in any part of the power system by the lightning stroke, this arc will set up very disturbing oscillations in the line. This may damage other equipment connected to the line.

6.5 PROTECTION AGAINST LIGHTNING

Transients or surges on the power system may originate from switching and from other causes but the most important and dangerous surges are those caused by lightning. The lightning surges may cause serious damage to the expensive equipment in the power system (e.g. generators, transformers etc.) either by direct strokes on the equipment or by strokes on the transmission lines that reach the equipment as travelling waves. It is necessary to provide protection against both kinds of surges. The most commonly used devices for protection against lightning surges are :

1. Earthing screen
2. Overhead ground wires
3. Lightning arresters or surge diverters

Earthing screen provides protection to power stations and sub-stations against direct strokes whereas overhead ground wires protect the transmission lines against direct lightning strokes. However, lightning arresters or surge diverters protect the station apparatus against both direct strokes and the strokes that come into the apparatus as travelling waves. We shall briefly discuss these methods of protection.

1. The Earthing Screen

- The power stations and sub-stations generally house expensive equipment. These stations can be protected against direct lightning strokes by providing earthing screen. It consists of a network of copper conductors (generally called shield or screen) mounted all over the electrical equipment in the sub-station or power station. The shield is properly connected to earth on atleast two points through a low impedance.

- On the occurrence of direct stroke on the station, screen provides a low resistance path by which lightning surges are conducted to ground. In this way, station equipment is protected against damage. The limitation of this method is that it does not provide protection against the travelling waves which may reach the equipment in the station.

2. Overhead Ground Wires

- The most effective method of providing protection to transmission lines against direct lightning strokes is by the use of overhead ground wires as shown in Fig. 6.10. For simplicity, one ground wire and one line conductor are shown. The ground wires are placed above the line conductors at such positions that practically all lightning strokes are intercepted by them (i.e. ground wires). The ground wires are grounded at each tower or pole through as low resistance as possible. Due to their proper location, the ground wires will take up all the lightning strokes instead of allowing them to line conductors.

- When the direct lightning stroke occurs on the transmission line, it will be taken up by the ground wires. The heavy lightning current (10 kA to 50 kA) from the ground wire flows to the ground, thus protecting the line from the harmful effects of lightning. It may be mentioned here that the degree of protection provided by the ground wires depends upon the footing resistance of the tower. Suppose, for example, tower-footing resistance is R_1 ohms and that the lightning current from tower to ground is I_1 amperes. Then the tower rises to a potential V_t given by ;

$$V_t = I_1 R_1$$

- Since V_t (= I_1R_1) is the approximate voltage between tower and line conductor, this is also the voltage that will appear across the string of insulators. If the value of V_t is less than that required to cause insulator flashover, no trouble results. On the other hand, if V_t is excessive, the insulator flashover may occur. Since the value of V_t depends upon tower-footing resistance R_1, the value of this resistance must be kept as low as possible to avoid insulator flashover.

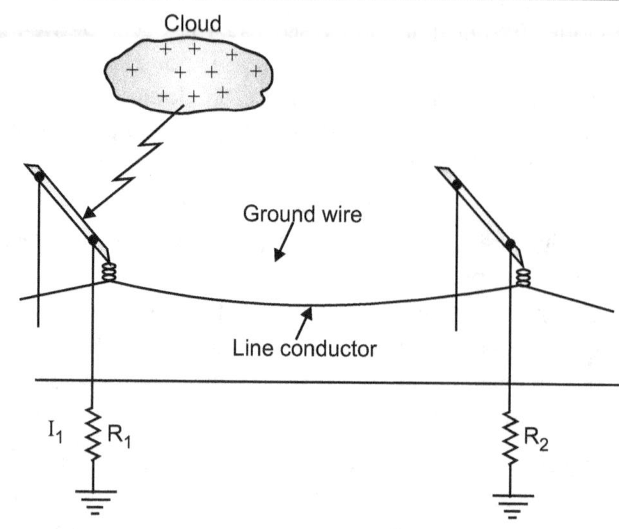

Fig. 6.10

Advantages

- It provides considerable protection against direct lightning strokes on transmission lines.

- A grounding wire provides damping effect on any disturbance travelling along the line as it acts as a short-circuited secondary.

- It provides a certain amount of electrostatic shielding against external fields. Thus it reduces the voltages induced in the line conductors due to the discharge of a neighbouring cloud.

Disadvantages

- It requires additional cost.

- There is a possibility of its breaking and falling across the line conductors, thereby causing a short-circuit fault. This objection has been greatly eliminated by using galvanised stranded steel conductors as ground wires. This provides sufficient strength to the ground wires.

6.6 LIGHTNING ARRESTERS

- The earthing screen and ground wires can well protect the electrical system against direct lightning strokes but they fail to provide protection against travelling waves which may reach the terminal apparatus. The lightning arresters or surge diverters provide protection against such surges.

- A lightning arrester or a surge diverter is a protective device which conducts the high voltage surges on the power system to the ground.

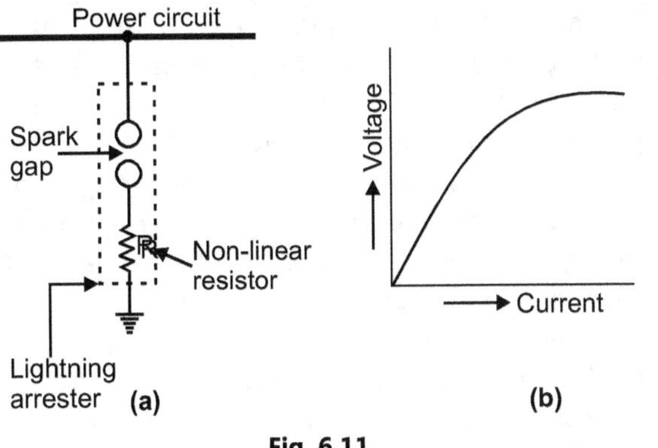

Fig. 6.11

- Fig. 6.11 (a) shows the basic form of a surge diverter. It consists of a spark gap in series with a non-linear resistor. One end of the diverter is connected to the terminal of the equipment to be protected and the other end is effectively grounded.

- The length of the gap is so set that normal line voltage is not enough to cause an arc across the gap but a dangerously high voltage will break down the air insulation and form an arc. The property of the non-linear resistance is that its resistance decreases as the voltage (or current) increases and vice-versa. This is clear from the volt/amp characteristic of the resistor shown in Fig. 6.11 (b).

Action : The action of the lightning arrester or surge diverter is as under :

(i) Under normal operation, the lightning arrester is off the line i.e. it conducts no current to earth or the gap is non-conducting.

(ii) On the occurrence of overvoltage, the air insulation across the gap breaks down and an arc is formed, providing a low resistance path for the surge to the ground. In this way, the excess charge on the line due to the surge is harmlessly conducted through the arrester to the ground instead of being sent back over the line.

(iii) It is worthwhile to mention the function of non-linear resistor in the operation of arrester.

As the gap sparks over due to overvoltage, the arc would be a short-circuit on the power system and may cause power-follow current in the arrester. Since the characteristic of the resistor is to offer high resistance to high voltage (or current), it prevents the effect of a short-circuit. After the surge is over, the resistor offers high resistance to make the gap non-conducting

Two things must be taken care of in the design of a lightning arrester. Firstly, when the surge is over, the arc in gap should cease. If the arc does not go out, the current

would continue to flow through the resistor and both resistor and gap may be destroyed. Secondly, IR drop (where I is the surge current) across the arrester when carrying surge current should not exceed the breakdown strength of the insulation of the equipment to be protected.

6.6.1 Types of Lightning Arresters

There are several types of lightning arresters in general use. They differ only in constructional details but operate on the same principle viz. providing low resistance path for the surges to the ground. We shall discuss the following types of lightning arresters :

1. Rod gap arrester
2. Horn gap arrester
3. Multigap arrester
4. Expulsion type lightning arrester
5. Valve type lightning arrester

1. **Rod Gap Arrester :** It is a very simple type of diverter and consists of two 1·5 cm rods which are bent at right angles with a gap in between as shown in Fig. 6.12. One rod is connected to the line circuit and the other rod is connected to earth. The distance between gap and insulator (i.e. distance P) must not be less than one-third of the gap length so that the arc may not reach the insulator and damage it. Generally, the gap length is so adjusted that breakdown should occur at 80% of spark- over voltage in order to avoid cascading of very steep wave fronts across the insulators. The string of insulators for an overhead line on the bushing of transformer has frequently a rod gap across it. Fig. 6.12 shows the rod gap across the bushing of a transformer.

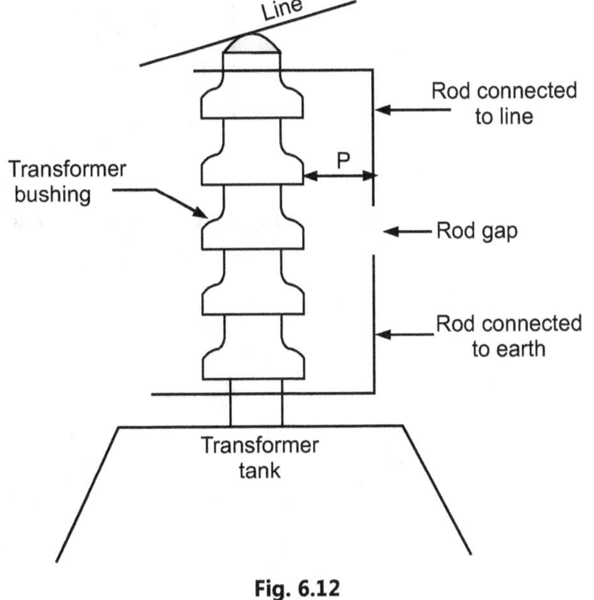

Fig. 6.12

Under normal operating conditions, the gap remains non-conducting. On the occurrence of a high voltage surge on the line, the gap sparks over and the surge current is conducted to earth. In this way, excess charge on the line due to the surge is harmlessly conducted to earth.

Limitations

- After the surge is over, the arc in the gap is maintained by the normal supply voltage, leading to a short-circuit on the system.

- The rods may melt or get damaged due to excessive heat produced by the arc.

- The climatic conditions (e.g. rain, humidity, temperature etc.) affect the performance of rod gap arrester.

- The polarity of the surge also affects the performance of this arrester.

- Due to the above limitations, the rod gap arrester is only used as a 'back-up' protection in case of main arresters.

2. Horn Gap Arrester

The horn gap consists of two horn shaped rods separated by a small distance. One end of this is connected to the line and the other to the earth as shown in Fig. 6.13, with or without a series resistance. The choke connected between the equipment to be protected and the horn gap serves two purposes:

(i) The steepness of the wave incident on the equipment to be protected is reduced.

(ii) It reflects the voltage surge back on to the horn.

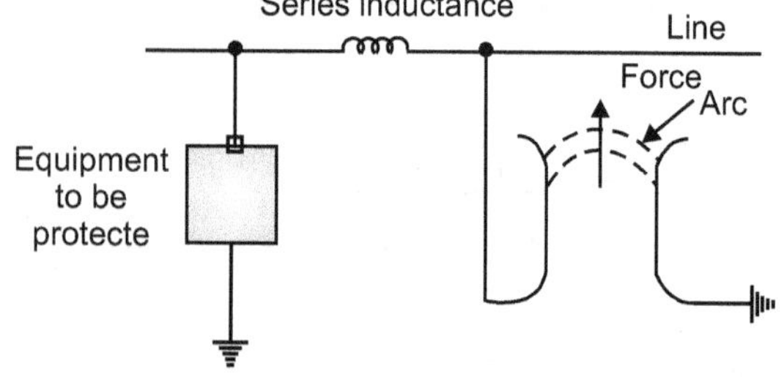

Fig. 6.13 : Horn gap connected in the system for protection

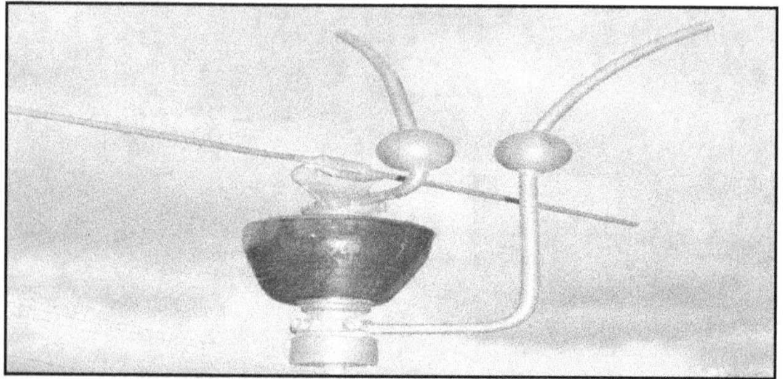

Fig. 6.14

- Whenever a surge voltage exceeds the breakdown value of the gap a discharge takes place and the energy content in the rest part of the wave is by-passed to the ground. An arc is set up between the gap, which acts like a flexible conductor and rises upwards under the influence of the electromagnetic forces, thus increasing the length of the arc which eventually blows out.

- There are two major drawbacks of the horn gap:
  (i) The time of operation of the gap is quite large as compared to the modern protective gear.
  (ii) If used on isolated neutral the horn gap may constitute a vicious kind of arcing ground. For these reasons, the horn gap has almost vanished from important power lines.

Advantages :

- The arc is self-clearing. Therefore, this type of arrester does not cause short-circuiting of the system after the surge is over as in the case of rod gap.

- Series resistance helps in limiting the follow current to a small value.

Limitations :

- The bridging of gap by some external agency (e.g. birds) can render the device useless.

- The setting of horn gap is likely to change due to corrosion or pitting. This adversely affects the performance of the arrester.

- The time of operation is comparatively long, say about 3 seconds. In view of the very short operating time of modern protective gear for feeders, this time is far long.

Due to the above limitations, this type of arrester is not reliable and can only be used as a second line of defence like the rod gap arrester.

3. **Multigap Arrester :** Fig. 6.15 shows the multigap arrester. It consists of a series of metallic (generally alloy of zinc) cylinders insulated from one another and separated by small intervals of air gaps. The first cylinder (i.e. A) in the series is connected to the line and the other to the ground through a series resistance. The series resistance limits the power arc. By the inclusion of series resistance, the degree of protection against travelling waves is reduced. In order to overcome this difficulty, some of the gaps (B to C in Fig. 6.15) are shunted by a resistance.

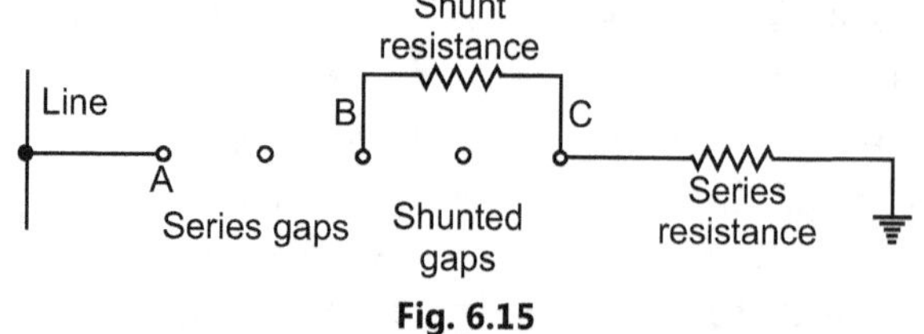

Fig. 6.15

- Under normal conditions, the point B is at earth potential and the normal supply voltage is unable to break down the series gaps. On the occurrence of an overvoltage, the breakdown of series gaps A to B occurs. The heavy current after breakdown will choose the straight - through path to earth via the shunted gaps B and C, instead of the alternative path through the shunt resistance. When the surge is over, the arcs B to C go out and any power current following the surge is limited by the two resistances (shunt resistance and series resistance) which are now in series. The current is too small to maintain the arcs in the gaps A to B and normal conditions are restored. Such arresters can be employed where system voltage does not exceed 33 kV.

4. Expulsion Type Arrester

- This type of arrester is also called 'protector tube' and is commonly used on system operating at voltages upto 33 kV. Fig. 6.16 (a) shows the essential parts of an expulsion type lightning arrester. It essentially consists of a rod gap A A' in series with a second gap enclosed within the fibre tube. The gap in the fibre tube is formed by two electrodes. The upper electrode is connected to rod gap and the lower electrode to the earth. One expulsion arrester is placed under each line conductor. Fig. 6.16 (b) shows the installation of expulsion arrester on an overhead line.

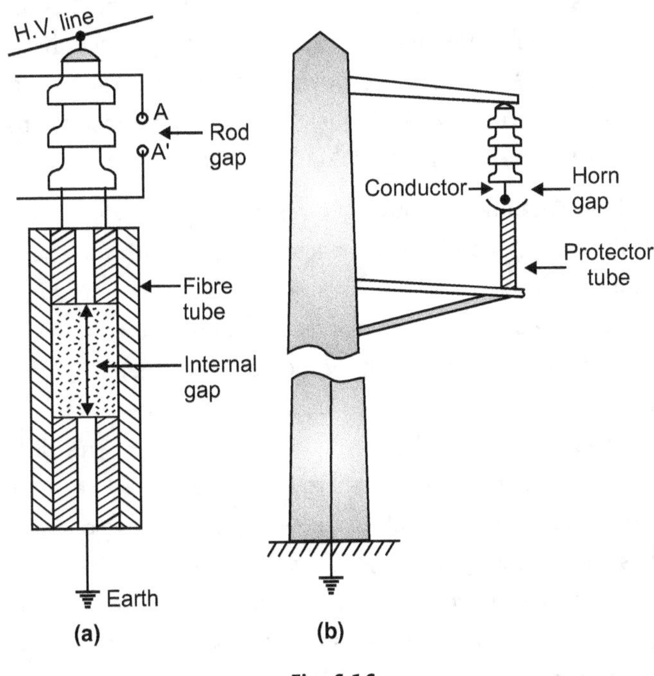

Fig. 6.16

On the occurrence of an overvoltage on the line, the series gap A A' is spanned and an arc is struck between the electrodes in the tube. The heat of the arc vaporises some of the fibre of tube walls, resulting in the production of a neutral gas. In an extremely short time, the gas builds up high pressure and is expelled through the lower electrode which is hollow. As the gas leaves the tube violently, it carries away ionised air around the arc. This de-ionising effect is generally so strong that arc goes out at a current zero and will not be re-established.

Advantages :

- They are not very expensive.

- They are improved form of rod gap arresters as they block the flow of power frequency follow currents.

- They can be easily installed

Limitations :

- An expulsion type arrester can perform only limited number of operations as during each operation some of the fibre material is used up.

- This type of arrester cannot be mounted in an enclosed equipment due to the discharge of gases during operation.

- Due to the poor volt/amp characteristic of the arrester, it is not suitable for the protection of expensive equipment.

5. **Valve Type Arrester :** Valve type arresters incorporate non-linear resistors and are extensively used on systems operating at high voltages. Fig. 6.17 (i) shows the various parts of a valve type arrester. It consists of two assemblies

   (i) series spark gaps and

   (ii) non-linear resistor discs (made of material such as thyrite or metrosil) in series. The non-linear elements are connected in series with the spark gaps. Both the assemblies are accommodated in tight porcelain container.

   (i) The spark gap is a multiple assembly consisting of a number of identical spark gaps in series. Each gap consists of two electrodes with a fixed gap spacing. The voltage distribution across the gaps is linearised by means of additional resistance elements (called grading resistors) across the gaps. The spacing of the series gaps is such that it will withstand the normal circuit voltage. However, an overvoltage will cause the gap to breakdown, causing the surge current to ground via the non-linear resistors.

(ii) The non-linear resistor discs are made of an inorganic compound such as Thyrite or Metrosil. These discs are connected in series. The non-linear resistors have the property of offering a high resistance to current flow when normal system voltage is applied, but a low resistance to the flow of high-surge currents. In other words, the resistance of these non-linear elements decreases with the increase in current through them and vice-versa.

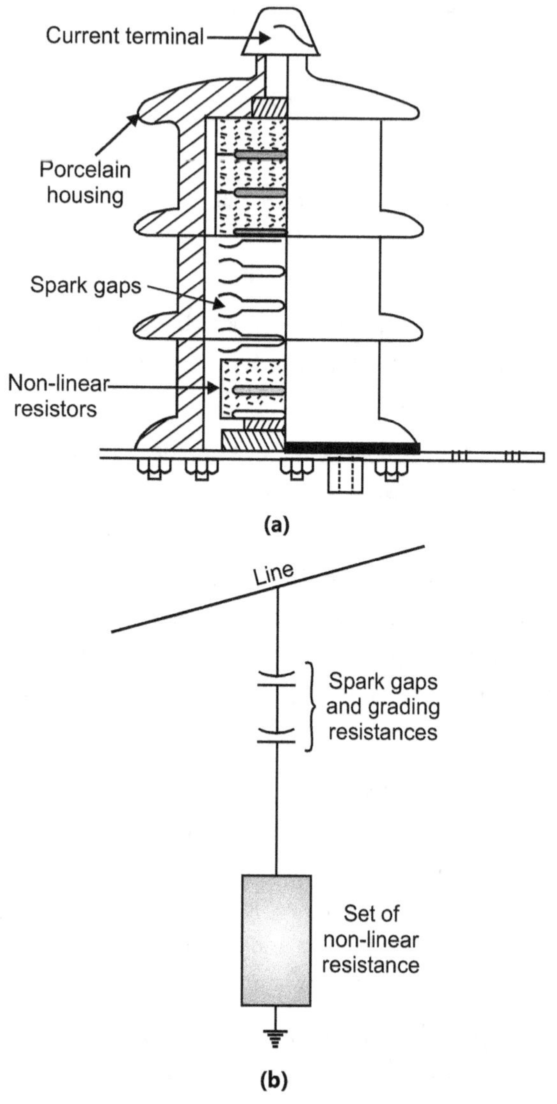

Fig. 6.17

Working : Under normal conditions, the normal system voltage is insufficient to cause the breakdown of air gap assembly. On the occurrence of an overvoltage, the breakdown of the series spark gap takes place and the surge current is conducted to earth via the non-linear resistors. Since the magnitude of surge current is very large, the non-linear elements will offer a very low resistance to the passage of surge. The result is that the surge will rapidly go to earth instead of being sent back over the line. When the surge is over, the non-linear resistors assume high resistance to stop the flow of current.

Advantages :

- They provide very effective protection (especially for transformers and cables) against surges.

- They operate very rapidly taking less than a second.

- The impulse ratio is practically unity.

Limitations

- They may fail to check the surges of very steep wave front from reaching the terminal apparatus. This calls for additional steps to check steep-fronted waves.

- Their performance is adversely affected by the entry of moisture into the enclosure. This necessitates effective sealing of the enclosure at all times.

6.6.2 Applications

According to their application, the valve type arresters are classified as (i) station type and (ii) line type. The station type arresters are generally used for the protection of important equipment in power stations operating on voltages upto 220 kV or higher. The line type arresters are also used for stations handling voltages upto 66 kV.

Silicon Carbide Lightening Arrester: A non linear Silicon Carbide (SiC) material is connected in series with the spark gaps as shown below. The spark gaps provide high impedance during normal condition, where as the SiC disks restricts the flow of current through the spark gap. The non linear resistor of SiC is made by mixing the same by binding material and forming a moulded disk. The disk diameter depends on its energy rating and thickness on the operating voltage rating. The V-I characteristics of a SiC has a hysteresis type loop, the resistance being high during the rising part of the impulse wave and it has a lower value during the tail of the wave front. These type of arresters are used upto a voltage level of 220 KV.

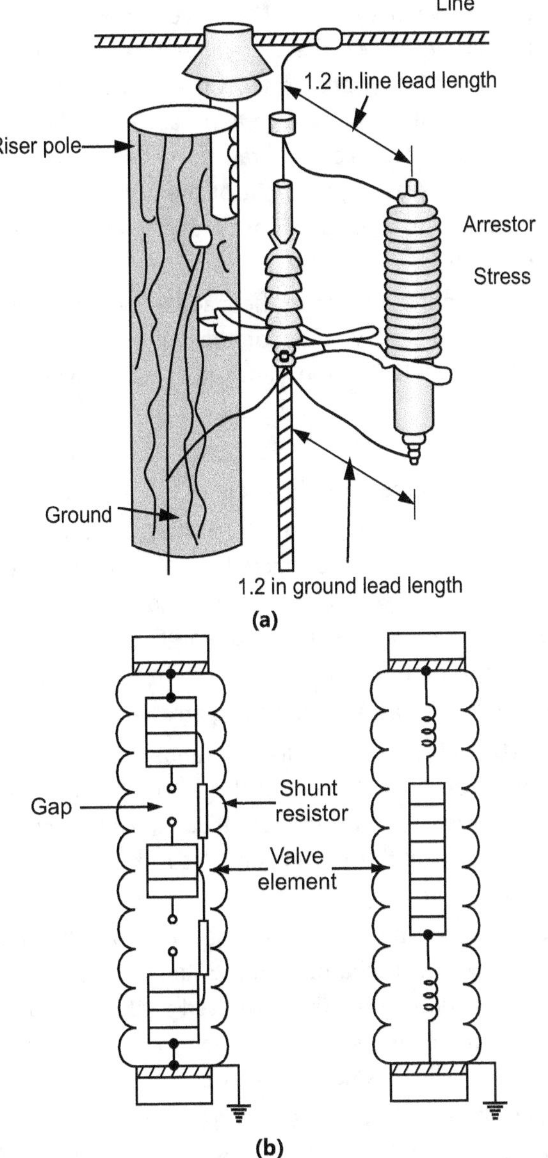

Fig. 6.18

Silicon Carbide Lighting Arrester: The Metal Oxide (MO) arrester shown below, uses metal oxide varistors for manufacturing. The main component being ZnO powder, mixed with some other metal oxides to form a ceramic mould. The characteristic is robust enough to avoid any use of spark gaps in series. Due to higher levels of leakage current compared to SiC type of arresters during transient overvoltage conditions. However, they carry even lesser current during normal voltage condition, therefore, they have even more non linear characteristics so that spark gaps can be avoided unlike the SiC type of arresters. During surge expulsion when the surge current is very high in the range of 250-500 A, a shunt gap provided in series with the arrester safeguards by bypassing with a spark over. These types of arresters are used for even more voltage levels.

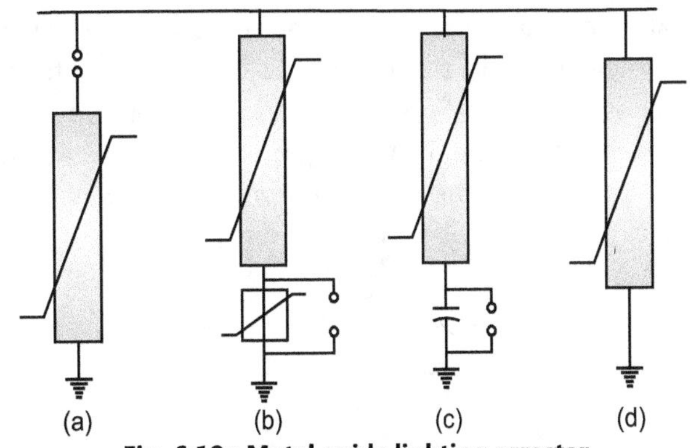

Fig. 6.19 : Metal oxide lighting arrester

6.7 SURGE ABSORBER

- The travelling waves set up on the transmission lines by the surges may reach the terminals apparatus and cause damage to it. The amount of damage caused not only depends upon the amplitude of the surge but also upon the steepness of its wave front. The steeper the wave front of the surge, the more the damage caused to the equipment. In order to reduce the steepness of the wave front of a surge, we generally use surge absorber.

- A surge absorber is a protective device which reduces the steepness of wave front of a surge by absorbing surge energy. Although both surge diverter and surge absorber eliminate the surge, the manner in which it is done is different in the two devices. The surge diverter diverts the surge to earth but the surge absorber absorbs the surge energy. A few cases of surge absorption are discussed below :

(i) A condenser connected between the line and earth can act as a surge absorber. Fig. 6.20 shows how a capacitor acts as surge absorber to protect the transformer winding. Since the reactance of a condenser is inversely proportional to frequency, it will be low at high frequency and high at low frequency. Since the surges are of high frequency, the capacitor acts as a short circuit and passes them directly to earth. However, for power frequency, the reactance of the capacitor is very high and practically no current flows to the ground.

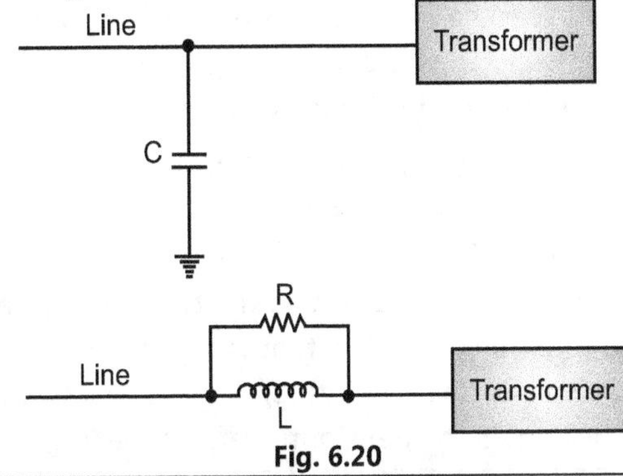

Fig. 6.20

(ii) Another type of surge absorber consists of a parallel combination of choke and resistance connected in series with the line as shown in Fig. 6.20. The choke offers high reactance to surge frequencies (XL = 2 π f L). The surges are, therefore, forced to flow through the resistance R where they are dissipated.

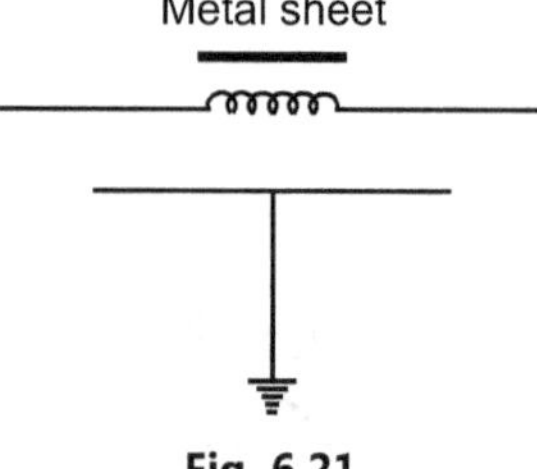

Fig. 6.21

(iii) Fig. 6.21 shows the another type of surge absorber. It is called Ferranti surge absorber. It consists of an air cored inductor connected in series with the line. The inductor is surrounded by but insulated from an earthed metallic sheet called dissipator. This arrangement is equivalent to a transformer with short-circuited secondary. The inductor forms the primary whereas the dissipator forms the short-circuited secondary. The energy of the surge is used up in the form of heat generated in the dissipator due to transformer action. This type of surge absorber is mainly used for the protection of transformers.

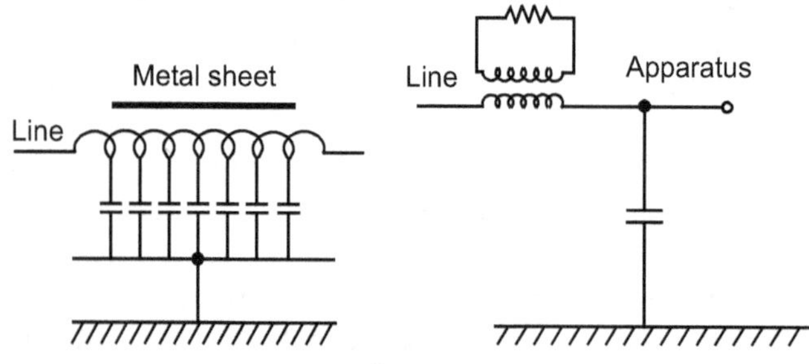

Fig. 6.22

Fig. 6.22 (a) shows the schematic diagram of 66 kV Ferranti surge absorber while Fig. 6.22 (b) shows its equivalent circuit.

6.8 LOCATION OF LIGHTNING ARRESTERS

The normal practice is to locate the lightning arrester as close as possible to the equipment to be protected. The following are the reasons for the practice:

- This reduces the chances of surges entering the circuit between the protective equipment and the equipment to be protected.

- If there is a distance between the two, a steep fronted wave after being incident on the lightning arrester, which sparks over corresponding to its spark-over voltage, enters the transformer after travelling over the lead between the two. The wave suffers reflection at the terminal and, therefore, the total voltage at the terminal of the transformer is the sum of reflected and the incident voltage which approaches nearly twice the incident voltage i.e., the transformer may experience a surge twice as high as that of the lightning arrester. If the lightning arrester is right at the terminals this could not occur.

- If L is the inductance of the lead between the two, and IR the residual voltage of the lightning arrester, the voltage incident at the transformer terminal will be

$$V = IR + L\frac{di}{dt}$$

Where di/dt is the rate of change of the surge current. It is possible to provide some separation between the two, where a capacitor is connected at the terminals of the equipment to be protected. This reduces the steepness of the wave and hence the rate di/dt and this also reduces the stress distribution over the winding of the equipment.

There are three classes of lightning arresters available:

(i) Station Type: The voltage ratings of such arresters vary from 3 kV to 312 kV and are designed to discharge currents not less than 100,000 amps. They are used for the protection of substation and power transformers.

(ii) Line Type: The voltage ratings vary from 20 kV to 73 kV and can discharge currents between 65,000 amps and 100,000 amps. They are used for the protection of distribution transformers, small power transformers and sometimes small substations.

(iii) Distribution Type: The voltage ratings vary from 8 kV to 15 kV and can discharge currents up to 65,000 amperes. They are used mainly for pole mounted substation for the protection of distribution transformers up to and including the 15 kV classification.

6.9 RATING OF LIGHTNING ARRESTERS

- A lightning arrester is expected to discharge surge currents of very large magnitude, thousands of amperes, but since the time is very short in terms of microseconds, the energy that is dissipated through the lightning arrester is small compared with what it would have been if a few amperes of power frequency current had been flown for a few cycles.

- Therefore, the main considerations in selecting the rating of a lightning arrester is the line-to-ground dynamic voltage to which the arrester may be subjected for any condition of system operation. An allowance of 5% is normally assumed, to take into account the light operating condition under no load at the far end of the line due to Ferranti effect and the sudden loss of load on water wheel generators. This means an arrester of 105% is used on a system where the line to ground voltage may reach line-to-line value during line-to-ground fault condition.

- The overvoltages on a system as is discussed earlier depend upon the neutral grounding condition which is determined by the parameters of the system. We recall that a system is said to be solidly grounded only if

$$\frac{R_0}{X_1} \leq 1$$

and

$$\frac{X_0}{X_1} \leq 3$$

and under this condition the line-to-ground voltage during a L-G fault does not exceed 80% of the L-L voltage and, therefore, an arrester of $(80\% + 0.05 \times 80\%) = 1.05 \times 80\% = 84\%$ is required. This is the extreme situation in case of solidly grounded system. In the same system the voltage may be less than 80%; say it may be 75%. In that case the rating of the lightning arrester will be $1.05 \times 75\% = 78.75\%$. The overvoltages can actually be obtained with the help of precalculated curves. One set of curves corresponding to a particular system is given in Fig. 6.23.

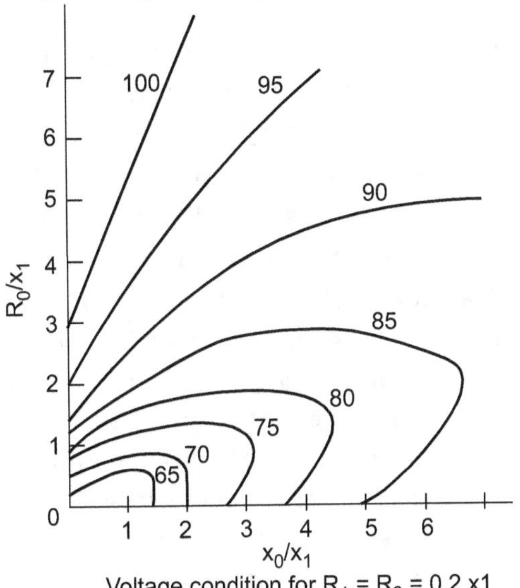

Fig. 6.23 : Maximum line-to-ground voltage at fault location for grounded neutral under any fault condition

For system grounded through Peterson coil, the overvoltages may be 100% if it is properly tuned and, therefore, it is customary to apply an arrester of 105% for such systems. Even though there is a risk of overvoltages becoming more than 100% if it is not properly tuned, but it is generally not feasible to select arresters of sufficiently high rating to eliminate all risks of arrester damage due to these reasons. The voltage rating of the arrester, therefore, ranges between 75% to 105% depending upon the neutral grounding condition.

So far we have discussed the non-shielding method. We now discuss the shielding method i.e., the use of ground wires for the protection of transmission lines against direct lightning strokes.

6.10 NEUTRAL GROUNDING

- In power system, grounding or earthing means connecting frame of electrical equipment (non-current carrying part) or some electrical part of the system (e.g. neutral point in a star-connected system, one conductor of the secondary of a transformer etc.) to earth i.e. soil. This connection to earth may be through a conductor or some other circuit element (e.g. a resistor, a circuit breaker etc.) depending upon the situation. Regardless of the method of connection to earth, grounding or earthing offers two principal advantages.

- First, it provides protection to the power system. For example, if the neutral point of a star-connected system is grounded through a circuit breaker and phase to earth fault occurs on any one line, a large fault current will flow through the circuit breaker. The circuit breaker will open to isolate the faulty line. This protects the power system from the harmful effects of the fault. Secondly, earthing of electrical equipment (e.g. domestic appliances, hand-held tools, industrial motors etc.) ensures the safety of the persons handling the equipment.

- For example, if insulation fails, there will be a direct contact of the live conductor with the metallic part (i.e. frame) of the equipment. Any person in contact with the metallic part of this equipment will be subjected to a dangerous electrical shock which can be fatal. In this chapter, we shall discuss the importance of grounding or earthing in the line of power system with special emphasis on neutral grounding.

6.10.1 Necessity of Power System Earthing

- **Overload Protection :** In scenarios where excessive power surge occurs, a grounded system helps immensely. This simple form of surge protection can instantly save your electrical appliances and devices from getting fried by excessive electrical power, saving your data as well as equipment.

- **Voltage Stabilization :** When it comes to calculating the right amount of power to be distributed between voltage sources, the earth provides that universal standard point of reference. Earthing takes the guesswork out of voltage stabilization, helping to ensure that no circuits overload or blow up.

- **Damage, Injury and Death Prevention :** Blown fuses or a tripped circuit breakers are far more welcome than electrical fires or shocks, which can pose serious safety hazards to people and property. Essentially, grounding protects against equipment, property and data loss, as well as injuries and fatalities!

6.10.2 Grounding or Earthing

- The process of connecting the metallic frame (i.e. non-current carrying part) of electrical equipment or some electrical part of the system (e.g. neutral point in a star-connected system, one conductor of the secondary of a transformer etc.) to earth (i.e. soil) is called **grounding** or **earthing.** It is strange but true that grounding of electrical systems is less understood aspect of power system.

- Nevertheless, it is a very important subject. If grounding is done systematically in the line of the power system, we can effectively prevent accidents and damage to the equipment of the power system and at the same time continuity of supply can be maintained.

- Grounding or earthing may be classified as :

  1. Equipment grounding

  2. System grounding.

- Equipment grounding deals with earthing the non-current-carrying metal parts of the electrical equipment. On the other hand, system grounding means earthing some part of the electrical system e.g. earthing of neutral point of star-connected system in generating stations and sub-stations.

6.10.3 Equipment Grounding

- The process of connecting non-current-carrying metal parts (i.e. metallic enclosure) of the electrical equipment to earth (i.e. soil) in such a way that in case of insulation failure, the enclosure effectively remains at earth potential is called equipment grounding.

- We are frequently in touch with electrical equipment of all kinds, ranging from domestic appliances and hand-held tools to industrial motors. We shall illustrate the need of effective equipment grounding by considering a single-phase circuit composed of a 230 V source

connected to a motor M as shown in Fig. 6.24. Note that neutral is solidly grounded at the service entrance.

- In the interest of easy understanding, we shall divide the discussion into three heads viz.

  (i)   Ungrounded enclosure

  (ii) Enclosure connected to neutral wire

  (iii) Ground wire connected to enclosure.

(i) Ungrounded Enclosure : Fig. 6.24 shows the case of ungrounded metal enclosure. If a person touches the metal enclosure, nothing will happen if the equipment is functioning correctly. But if the winding insulation becomes faulty, the resistance Re between the motor and enclosure drops to a low value (a few hundred ohms or less). A person having a body resistance Rb would complete the current path as shown in Fig. 6.24

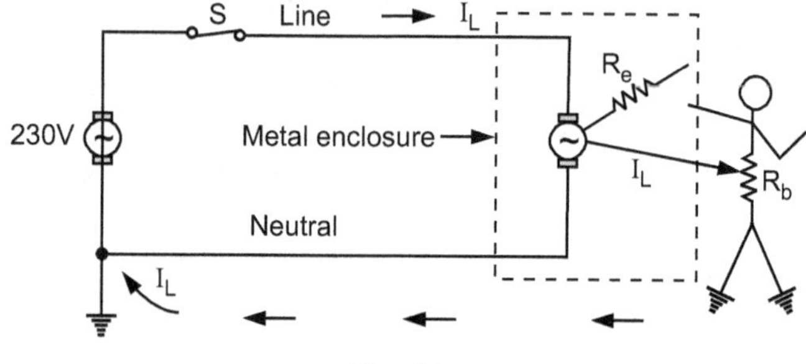

Fig. 6.24

If Re is small (as is usually the case when insulation failure of winding occurs), the leakage current IL through the person's body could be dangerously high. As a result, the person would get severe electric shock which may be fatal. Therefore, this system is unsafe.

(ii) Enclosure Connected to Neutral Wire : It may appear that the above problem can be solved by connecting the enclosure to the grounded neutral wire as shown in Fig. 6.25. Now the leakage current IL flows from the motor, through the enclosure and straight back to the neutral wire (See Fig. 6.25). Therefore, the enclosure remains at earth potential. Consequently, the operator would not experience any electric shock.

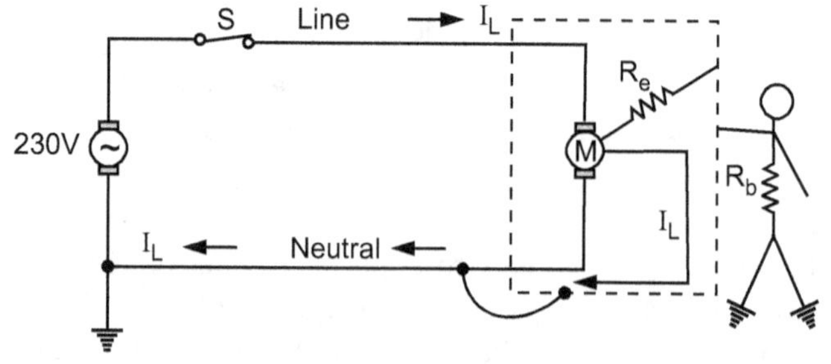

Fig. 6.25

The trouble with this method is that the neutral wire may become open either accidentally or due to a faulty installation. For example, if the switch is inadvertently in series with the neutral rather than the live wire (See Fig. 6.26), the motor can still be turned on and off. However, if someone touched the enclosure while the motor is off, he would receive a severe electric shock (See Fig. 6.26). It is because when the motor is off, the potential of the enclosure rises to that of the live conductor.

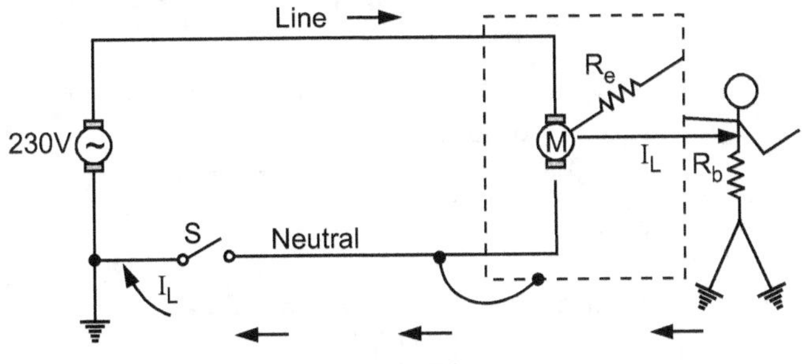

Fig. 6.26

(iii) Ground Wire Connected to Enclosure : To get rid of this problem, we install a third wire, called ground wire, between the enclosure and the system ground as shown in Fig. 6.27. The ground wire may be bare or insulated. If it is insulated, it is coloured green.

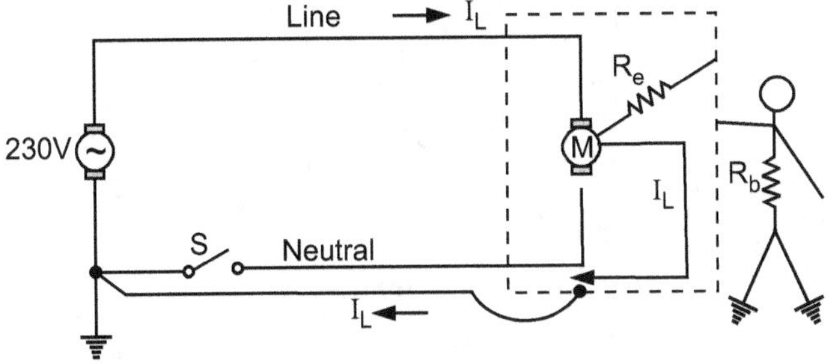

Fig. 6.27

Electrical outlets have three contacts — one for live wire, one for neutral wire and one for ground wire

6.10.4 System Grounding

- The process of connecting some electrical part of the power system (e.g. neutral point of a star connected system, one conductor of the secondary of a transformer etc.) to earth (i.e. soil) is called **system grounding**.

- The system grounding has assumed considerable importance in the fast expanding power system.

- By adopting proper schemes of system grounding, we can achieve many advantages including protection, reliability and safety to the power system network. But before discussing the various aspects of neutral grounding, it is desirable to give two examples to appreciate the need of system grounding.

(i) Fig. 6.28 (a) shows the primary winding of a distribution transformer connected between the line and neutral of a 11 kV line. If the secondary conductors are ungrounded, it would appear that a person could touch either secondary conductor without harm because there is no ground return. However, this is not true. Referring to Fig. 6.28, there is capacitance C_1 between primary and secondary and capacitance C_2 between secondary and ground. This capacitance coupling can produce a high voltage between the secondary lines and the ground. Depending upon the relative magnitudes of C_1 and C_2, it may be as high as 20% to 40% of the primary voltage. If a person touches either one of the secondary wires, the resulting capacitive current I_C flowing through the body could be dangerous even in case of small transformers [See Fig. 6.28 (b)]. For example, if I_C is only 20 mA, the person may get a fatal electric shock.

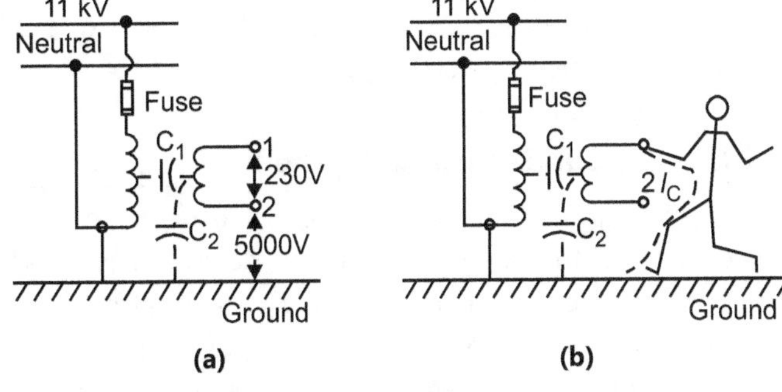

Fig. 6.28

If one of the secondary conductors is grounded, the capacitive coupling almost reduces to zero and so is the capacitive current I_C. As a result, the person will experience no electric shock. This explains the importance of system grounding.

(ii) Let us now turn to a more serious situation. Fig. 6.29 (a) shows the primary winding of a distribution transformer connected between the line and neutral of a 11 kV line. The secondary conductors are ungrounded. Suppose that the high voltage line (11 kV in this case) touches the 230 V conductor as shown in Fig. 6.28 (a). This could be caused by an internal fault in the transformer or by a branch or tree falling across the 11 kV and 230 V lines. Under these circumstances, a very high voltage is imposed between the secondary conductors and ground. This would immediately puncture the 230 V insulation, causing a massive flashover. This flashover could occur anywhere on the secondary network, possibly inside a home or factory. Therefore, ungrounded secondary in this case is a potential fire hazard and may produce grave accidents under abnormal conditions.

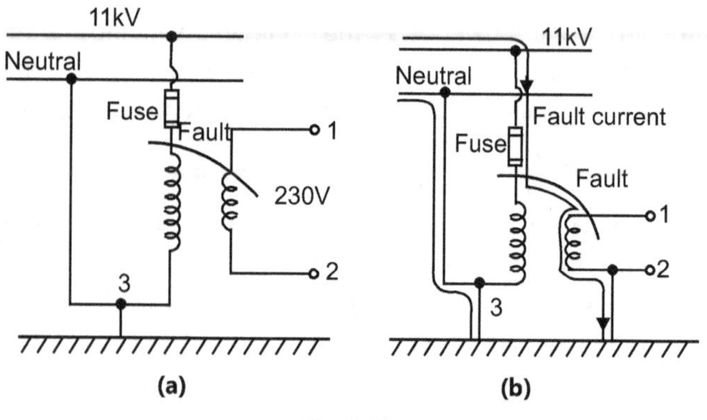

(a)   (b)

Fig. 6.29

- If one of the secondary lines is grounded as shown in Fig. 6.29 (b), the accidental contact between a 11 kV conductor and a 230 V conductor produces a dead short. The short-circuit current (i.e. fault current) follows the dotted path shown in Fig. 6.29 (b). This large current will blow the fuse on the 11 kV side, thus disconnecting the transformer and secondary distribution system from the 11 kV line. This explains the importance of system grounding in the line of the power system.

6.10.5 Ungrounded Neutral System

- In an ungrounded neutral system, the neutral is not connected to the ground i.e. the neutral is isolated from the ground. Therefore, this system is also called isolated neutral system or free neutral system. Fig. 6.30 shows ungrounded neutral system. The line conductors have capacitances between one another and to ground. The former are delta-connected while the latter are star-connected. The delta-connected capacitances have little effect on the grounding characteristics of the system (i.e. these capacitances do not effect the earth circuit) and, therefore, can be neglected. The circuit then reduces to the one shown in Fig. 6.29(i).

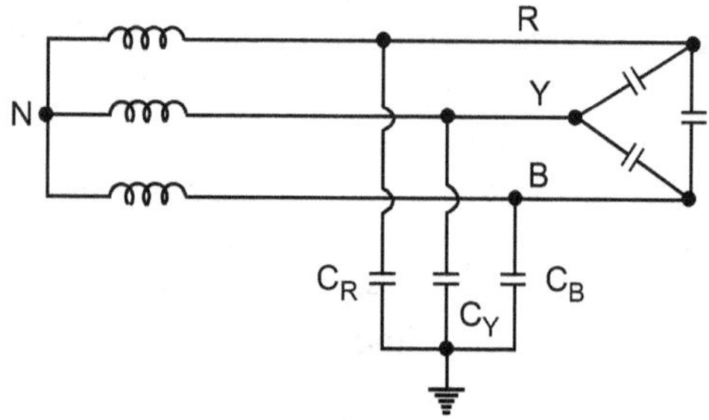

Fig. 6.30

6.10.6 Circuit Behavior Under Normal Conditions

- Let us discuss the behaviour of ungrounded neutral system under normal conditions (i.e. under steady state and balanced conditions). The line is assumed to be perfectly transposed so that each conductor has the same capacitance to ground.

Therefore, $C_R = C_Y = C_B = C$ (say). Since the phase voltages V_{RN}, V_Y N and V_{BN} have the same magnitude (of course, displaced 120° from one another), the capacitive currents I_R, I_Y and I_B will have the same value i.e.

$$I_R = I_Y = I_B = \frac{V_{ph}}{X_C} \quad \text{...in magnitude}$$

where V_{ph} = Phase voltage (i.e. line-to-neutral voltage)

            X_C = Capacitive reactance of the line to ground

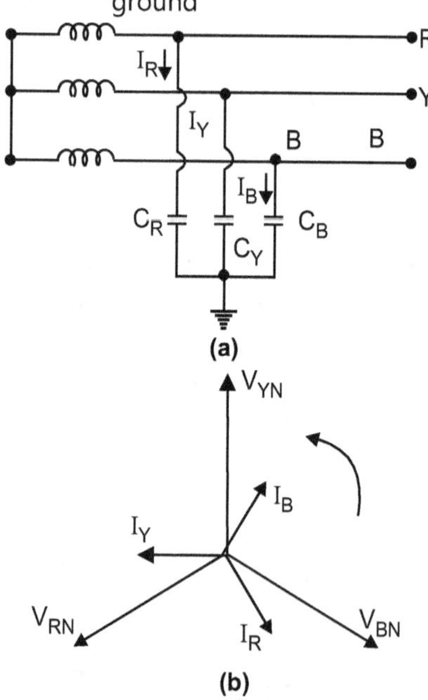

(a)

(b)

Fig. 6.31

The capacitive currents I_R, I_Y and I_B lead their respective phase voltages V_{RN}, V_Y N and V_{BN} by 90° as shown in the phasor diagram in Fig. 6.31 (b). The three capacitive currents are equal in magnitude and are displaced 120° from each other. Therefore, their phasor sum is zero. As a result, no current flows to ground and the potential of neutral is the same as the ground potential. Therefore, ungrounded neutral system poses no problems under normal conditions. However, as we shall see, currents and voltages are greatly influenced during fault conditions.

6.10.7 Circuit Behaviour Under Single Line to Ground-Fault

Let us discuss the behaviour of ungrounded neutral system when single line to ground fault occurs. Suppose line to ground fault occurs in line B at some point F. The circuit then becomes as shown in Fig. 6.32 (a). The capacitive currents I_R and I_Y flow through the lines R and Y respectively. The voltages driving IR and I_Y are V_{BR} and V_{BY} respectively. Note that VBR and V_B Y are the line voltages [See Fig. 6.32 (b)]. The paths of I_R and I_Y are essentially capacitive. Therefore, I_R leads V_{BR} by 90° and I_Y leads V_B Y by 90° as shown in Fig. 6.32 (b). The capacitive fault current I_C in line B is the phasor sum of I_R and IY .

Fault current in line B,

$$I_C = I_R + I_Y \qquad \text{...Phasor sum}$$

Now, $\qquad I_R = \dfrac{V_{BR}}{X_C} = \dfrac{\sqrt{3}\,V_{ph}}{X_C}$

and $\qquad I_Y = \dfrac{V_{BY}}{X_C} = \dfrac{\sqrt{3}\,V_{ph}}{X_C}$

$\therefore \qquad I_R = I_Y = \dfrac{\sqrt{3}\,V_{ph}}{X_C}$

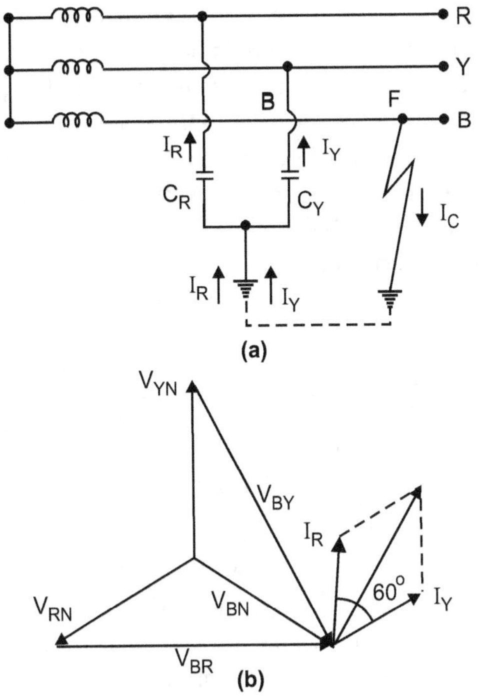

Fig. 6.32

$$= \sqrt{3} \times \text{ Per phase capacitive current under normal conditions}$$

Capacitive fault current in line B is

$$I_C = \text{Phasor sum of } I_R \text{ and } I_Y$$

$$= \sqrt{3}\,I_R = \sqrt{3} \times \dfrac{\sqrt{3}\,V_{ph}}{X_C} = \dfrac{3V_{ph}}{X_C}$$

$$\therefore \qquad I_C = \dfrac{3V_{ph}}{X_C} = 3 \times \dfrac{V_{ph}}{X_C}$$

$$= 3 \times \text{ per phase capacitive current under normal conditions}$$

Therefore, when single line to ground fault occurs on an ungrounded neutral system, the following effects are produced in the system:

- The potential of the faulty phase becomes equal to ground potential. However, the voltages of the two remaining healthy phases rise from their normal phase voltages to full line value. This may result in insulation breakdown.

- The capacitive current in the two healthy phases increase to 3 times the normal value.

- The capacitive fault current (I_C) becomes 3 times the normal per phase capacitive current.

- This system cannot provide adequate protection against earth faults. It is because the capacitive fault current is small in magnitude and cannot operate protective devices.

- The capacitive fault current I_C flows into earth. Experience shows that I_C in excess of 4A is sufficient to maintain an arc in the ionized path of the fault. If this current is once maintained, it may exist even after the earth fault is cleared. This phenomenon of persistent arc is called arcing ground. Due to arcing ground, the system capacity is charged and discharged in a cyclic order. This sets up high-frequency oscillations on the whole system and the phase voltage of healthy conductors may rise to 5 to 6 times its normal value. The overvoltages in healthy conductors may damage the insulation in the line.

Due to above disadvantages, ungrounded neutral system is not used these days. The modern high-voltage 3-phase systems employ grounded neutral owing to a number of advantages.

Neutral Grounding

- The process of connecting neutral point of 3-phase system to earth (i.e. soil) either directly or through some circuit element (e.g. resistance, reactance etc.) is called neutral grounding.

- Neutral grounding provides protection to personal and equipment. It is because during earth fault, the current path is completed through the earthed neutral and the protective devices (e.g. a fuse etc.) operate to isolate the faulty conductor from the rest of the system. This point is illustrated in Fig. 6.33.

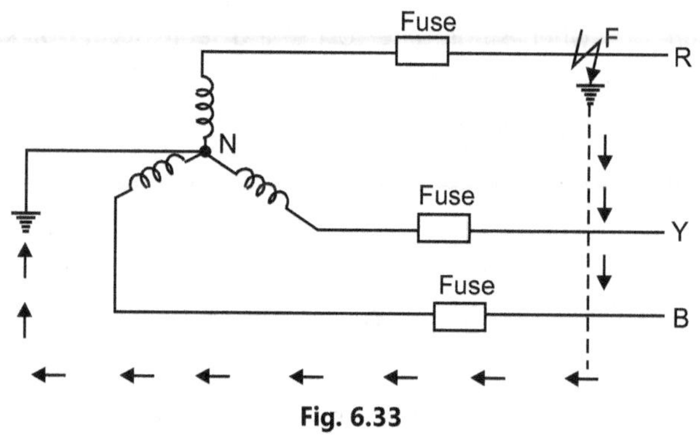

Fig. 6.33

- Fig. 6.33 shows a 3-phase, star-connected system with neutral earthed (i.e. neutral point is connected to soil). Suppose a single line to ground fault occurs in line R at point F. This will cause the current to flow through ground path as shown in Fig. 6.33. Note that current flows from R phase to earth, then to neutral point N and back to R-phase.

- Since the impedance of the current path is low, a large current flows through this path. This large current will blow the fuse in R-phase and isolate the faulty line R. This will protect the system from the harmful effects (e.g. damage to equipment, electric shock to personnel etc.) of the fault. One important feature of grounded neutral is that the potential difference between the live conductor and ground will not exceed the phase voltage of the system i.e. it will remain nearly constant.

Advantages of Neutral Grounding

The following are the advantages of neutral grounding:

- Voltages of the healthy phases do not exceed line to ground voltages i.e. they remain nearly constant.

- The high voltages due to arcing grounds are eliminated.

- The protective relays can be used to provide protection against earth faults. In case earth fault occurs on any line, the protective relay will operate to isolate the faulty line.

- The overvoltages due to lightning are discharged to earth.

- It provides greater safety to personnel and equipment.

- It provides improved service reliability.

- Operating and maintenance expenditures are reduced.

6.11 METHODS OF NEUTRAL GROUNDING

The methods commonly used for grounding the neutral point of a 3-phase system are :

1. Solid or effective grounding
2. Resistance grounding
3. Reactance grounding
4. Peterson-coil grounding

The choice of the method of grounding depends upon many factors including the size of the system, system voltage and the scheme of protection to be used.

1. Solid Grounding

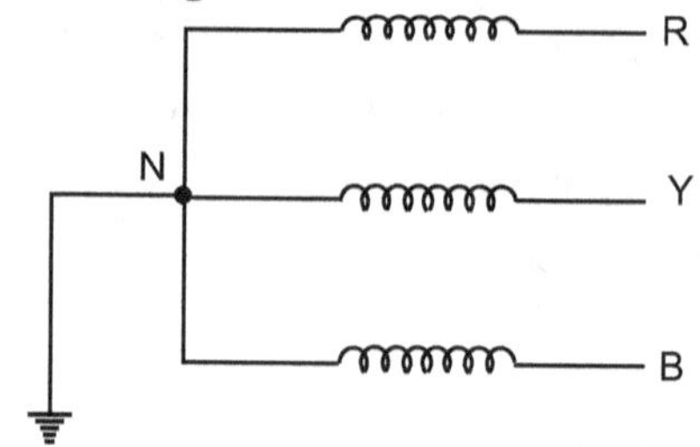

Fig. 6.34

- When the neutral point of a 3-phase system (e.g. 3-phase generator, 3-phase transformer etc.) is directly connected to earth (i.e. soil) through a wire of negligible resistance and reactance, it is called solid grounding or effective grounding. Fig. 6.34 shows the solid grounding of the neutral point. Since the neutral point is directly connected to earth through a wire, the neutral point is held at earth potential under all conditions. Therefore, under fault conditions, the voltage of any conductor to earth will not exceed the normal phase voltage of the system.

2. Resistance Grounding

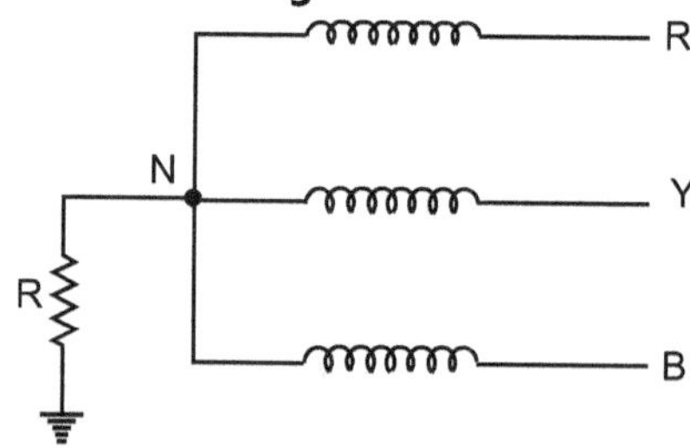

Fig. 6.35

- In order to limit the magnitude of earth fault current, it is a common practice to connect the neutral point of a 3-phase system to earth through a resistor. This is called resistance grounding. When the neutral point of a 3-phase system (e.g. 3-phase generator, 3-phase transformer etc.) is connected to earth (i.e. soil) through a resistor, it is called resistance grounding. Fig. 6.35 shows the grounding of neutral point through a resistor R. The value of R should neither be very low nor very high.

- If the value of earthing resistance R is very low, the earth fault current will be large and the system becomes similar to the solid grounding system. On the other hand, if the earthing resistance R is very high, the system conditions become similar to ungrounded neutral system.

- The value of R is so chosen such that the earth fault current is limited to safe value but still sufficient to permit the operation of earth fault protection system. In practice, that value of R is selected that limits the earth fault current to 2 times the normal full load current of the earthed generator or transformer.

3. Reactance Grounding

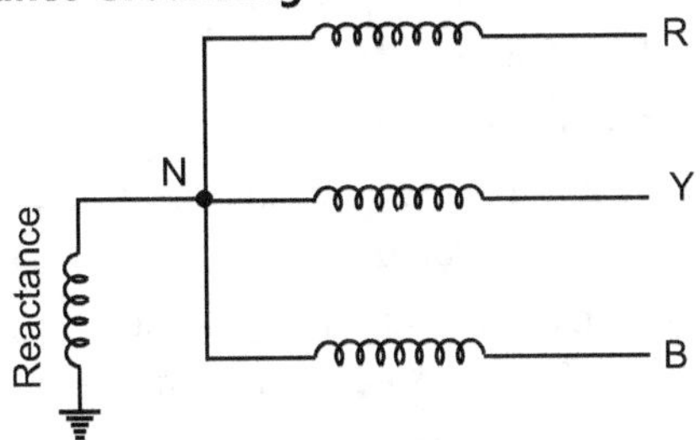

Fig. 6.36

- In this system, a reactance is inserted between the neutral and ground as shown in Fig. 6.36. The purpose of reactance is to limit the earth fault current. By changing the earthing reactance, the earth fault current can to changed to obtain the conditions similar to that of solid grounding. This method is not used these days because of the following disadvantages :

- In this system, the fault current required to operate the protective device is higher than that of resistance grounding for the same fault conditions.

- High transient voltages appear under fault conditions.

4. Arc Suppression Coil Grounding (or Resonant Grounding or Peterson coil grounding)

- We have seen that capacitive currents are responsible for producing arcing grounds. These capacitive currents flow because capacitance exists between each line and earth. If inductance L of appropriate value is connected in parallel with the capacitance of the system, the fault current I_F flowing through L will be in phase opposition to the capacitive current I_C of the system. If L is so adjusted that $I_L = I_C$, then resultant current in the fault will be zero. This condition is known as resonant grounding. When the value of L of arc suppression coil is such that the fault current IF exactly balances the capacitive current I_C, it is called resonant grounding.

- Circuit details An arc suppression coil (also called Peterson coil) is an iron-cored coil connected between the neutral and earth as shown in Fig. 6.37 (a). The reactor is provided with tappings to change the inductance of the coil. By adjusting the tappings on the coil, the coil can be tuned with the capacitance of the system i.e. resonant grounding can be achieved.

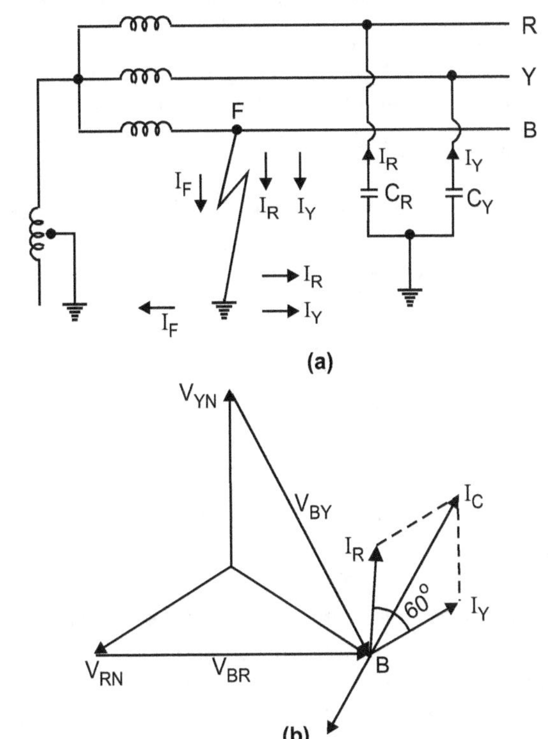

Fig. 6.37

Operation : Fig. 6.37 (a) shows the 3-phase system employing Peterson coil grounding. Suppose line to ground fault occurs in the line B at point F. The fault current IF and capacitive currents IR and IY will flow as shown in Fig. 6.37 (a). Note that IF flows through the Peterson coil (or Arc suppression coil) to neutral and back through the fault. The total capacitive current IC is the phasor sum of I_R and I_Y as shown in phasor diagram in Fig. 6.37 (b). The voltage of the faulty phase is applied across the arc suppression coil. Therefore, fault current I_F lags the faulty phase voltage by 90°. The current IF is in phase opposition to capacitive current I_C [See Fig. 6.37 (b)]. By adjusting the tappings on the Peterson coil, the resultant current in the fault can be reduced. If inductance of the coil is so adjusted that $I_L = I_C$, then resultant current in the fault will be zero.

Value of L for Resonant Grounding : For resonant grounding, the system behaves as an ungrounded neutral system. Therefore, full line voltage appears across capacitors C_R and C_Y.

$$\therefore \quad I_R = I_Y = \frac{\sqrt{3}\, V_{ph}}{X_C}$$

$$\therefore \quad I_C = \sqrt{3}\, I_R = \sqrt{3} \times \frac{\sqrt{3}\, V_{ph}}{X_C} = \frac{3 V_{ph}}{X_C}$$

Here, X_C is the line to ground capacitive reactance.

Fault current, $I_F = \dfrac{V_{ph}}{X_L}$

Here, X_L is the line to ground capacitive reactance.

Fault current, $I_F = \dfrac{V_{ph}}{X_L}$

Here, X_L is the inductive reactance of the arc suppression coil.

For resonant grounding,

$$I_L = I_C$$

or $$\dfrac{V_{ph}}{X_L} = \dfrac{3V_{ph}}{X_C}$$

or $$X_L = \dfrac{X_C}{3}$$

or $$\omega L = \dfrac{1}{3\omega C}$$

$\therefore$ $$L = \dfrac{1}{3\omega^2 C} \qquad \ldots(6.1)$$

Exp. (6.1) gives the value of inductance L of the arc suppression coil for resonant grounding.

Advantages : The Peterson coil grounding has the following advantages:

- The Peterson coil is completely effective in preventing any damage by an arcing ground.
- The Peterson coil has the advantages of ungrounded neutral system.

Disadvantages : The Peterson coil grounding has the following disadvantages :

- Due to varying operational conditions, the capacitance of the network changes from time to time. Therefore, inductance L of Peterson coil requires readjustment.
- The lines should be transposed.

6.12 VOLTAGE TRANSFORMER EARTHING

In this method of neutral earthing, the primary of a single-phase voltage transformer is connected between the neutral and the earth as shown in Fig. 6.38. A low resistor in series with a relay is connected across the secondary of the voltage transformer. The voltage transformer provides a high reactance in the neutral earthing circuit and operates virtually as an ungrounded neutral system. An earth fault on any phase produces a voltage across the relay. This causes the operation of the protective device.

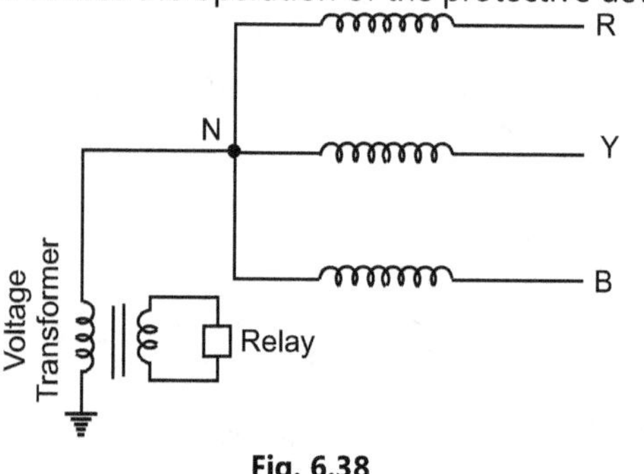

Fig. 6.38

Advantages : The following are the advantages of voltage transformer earthing :

- The transient overvoltages on the system due to switching and arcing grounds are reduced.

  It is because voltage transformer provides high reactance to the earth path.

- This type of earthing has all the advantages of ungrounded neutral system.
- Arcing grounds are eliminated.

Disadvantages : The following are the disadvantages of voltage transformer earthing :

- When earth fault occurs on any phase, the line voltage appears across line to earth capacitances. The system insulation will be overstressed.
- The earthed neutral acts as a reflection point for the raveling waves through the machine winding. This may result in high voltage build up.

Applications : The use of this system of neutral earthing is normally confined to generator equipments which are directly connected to step-up power transformers.

SOLVED EXAMPLES

Example 6.1 : *Calculate the reactance of Peterson coil suitable for a 33 kV, 3-phase transmission line having a capacitance to earth of each conductor as 4.5 μF. Assume supply frequency to be 50 Hz.*

Solution : Supply frequency, f = 50 Hz

Line to earth capacitance , $C = 4.5\ \mu F = 4.5 \times 10^{-6}\ F$

For Peterson coil grounding, reactance X_L of the Peterson coil should be equal to $\dfrac{X_C}{3}$ where X_C is line to earth capacitive reactance.

$\therefore$ Reactance of Peterson coil,

$$X_L = \dfrac{X_C}{3} = \dfrac{1}{3\omega C} = \dfrac{1}{3 \times 2\pi f \times C}$$

$$= \dfrac{1}{3 \times 2\pi \times 50 \times 4.5 \times 10^{-6}} = 235.8\ \Omega$$

Example 6.2: *A 230 kV, 3-phase, 50 Hz, 200 km transmission line has a capacitance to earth of 0.02 μF/km per phase. Calculate the inductance and kVA rating of the Peterson coil used for earthing the above system.*

Solution : Supply frequency ,

$$f = 50\ Hz$$

Capacitance of each line to earth,

$$C = 200 \times 0.02 = 4 \times 10^{-6}\ F$$

Required inductance of Peterson coil is

$$L = \frac{1}{3\omega^2 C}$$

$$= \frac{1}{3 \times (2\pi \times 50)^2 \times 4 \times 10^{-6}} = 0.85 \text{ H}$$

Current through Peterson coil is

$$I_F = \frac{V_{ph}}{X_L} = \frac{230 \times 10^3/\sqrt{3}}{2\pi \times 50 \times 0.85} = 500 \text{ A}$$

Voltage across Peterson coil is

$$V_{ph} = \frac{V_L}{\sqrt{3}} = \frac{230 \times 1000}{\sqrt{3}} \text{ V}$$

$\therefore$ Rating of Peterson coil

$$= V_{ph} \times I_F$$

$$= \frac{230 \times 1000}{\sqrt{3}} \times 500 \times \frac{1}{1000} \text{ kVA}$$

$$= 66397 \text{ kVA}$$

Example 6.3 : *A 50 Hz overhead line has line to earth capacitance of 1.2 μF. It is desired to use earth fault neutralizer. Determine the reactance to neutralize the capacitance of (i) 100% of the length of the line (ii) 90% of the length of the line and (iii) 80% of the length of the line.*

Solution :

(i) Inductive reactance of the coil to neutralize capacitance of 100% of the length of the line is

$$X_L = \frac{1}{3\omega C} = \frac{1}{3 \times 2\pi \times 50 \times 1.2 \times 10^{-6}}$$

$$= 884.19 \ \Omega$$

(ii) Inductive reactance of the coil to neutralize capacitance of 90% of the length of the line is

$$X_L = \frac{1}{3\omega \times 0.9 C}$$

$$= \frac{1}{3 \times 2\pi \times 50 \times 0.9 \times 1.2 \times 10^{-6}}$$

$$= 982.43 \ \Omega$$

(iii) Inductive reactance of the coil to neutralize capacitance of 80% of the length of the line is

$$X_L = \frac{1}{3\omega \times 0.8 \ C}$$

$$= \frac{1}{3 \times 2\pi \times 50 \times 0.8 \times 1.2 \times 10^{-6}}$$

$$= 1105.24 \ \Omega$$

Example 6.4 : *Calculate the reactance of Peterson coil suitable for a 33 kV, 3-phase transmission line having a capacitance to earth of each conductor as 4.5 μF. Assume supply frequency to be 50 Hz.*

Solution : Supply frequency, f = 50 Hz

Line to earth capacitance, c = 4.5 μF = 4.5×10^{-6} F

For Peterson coil grounding, reactance X_L of the Peterson coil should be equal to $X_C/3$ where X_C is line to earth capacitive reactance.

$\therefore$ Reactance of Peterson coil,

$$X_L = \frac{X_C}{3} = \frac{1}{3\omega C} = \frac{1}{3 \times 2\pi f \times C}$$

$$= \frac{1}{3 \times 2\pi \times 50 \times 4.5 \times 10^{-6}} = 235.8 \ \Omega$$

Example 6.5 : *A 230 kV, 3-phase, 50 Hz, 200 km transmission line has a capacitance to earth of 0.02 μF/km per phase. Calculate the inductance and kVA rating of the Peterson coil used for earthing the above system.*

Solution : Supply frequency , f = 50 Hz

Capacitance of each line to earth,

$$C = 200 \times 0.02 = 4 \times 10^{-6} \text{ F}$$

Required inductance of Peterson coil is

6.12.1 Grounding Transformer

- We sometimes have to create a neutral point on a 3-phase, 3-wire system (e.g. delta connection etc.) to change it into 3-phase, 4-wire system. This can be done by means of a grounding transformer. It is a core type transformer having three limbs built in the same fashion as that of the power transformer.

- Each limb of the transformer has two identical windings wound differentially (i.e. directions of current in the two windings on each limb are opposite to each other) as shown in Fig. 6.39. Under normal operating conditions, the total flux in each limb is negligibly small. Therefore, the transformer draws very small magnetising current.

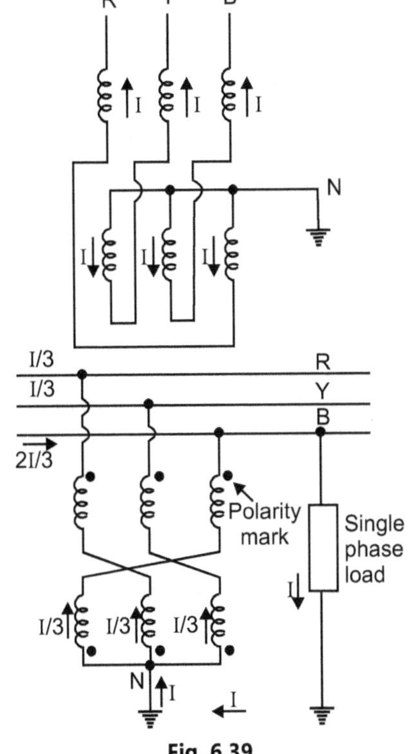

Fig. 6.39

- Fig. 6.39 shows the use of grounding transformer to create neutral point N. If we connect a single-phase load between one line and neutral, the load current I divides into three equal currents in each winding. Because the currents are equal, the neutral point stays fixed and the line to neutral voltages remain balanced as they would be on a regular 4-wire system. In practice, the single-phase loads are distributed as evenly as possible between the three phases and neutral so that unbalanced load current I is relatively small.

- The impedance of grounding transformer is quite low. Therefore, when line to earth fault occurs, the fault current will be quite high. The magnitude of fault current is limited by inserting a resistance (not shown in the figure) in the neutral circuit. Under normal conditions, only iron losses will be continuously occurring in the grounding transformer. However, in case of fault, the high fault current will also produce copper losses in the transformer. Since the duration of the fault current is generally between 30-60 seconds, the copper losses will occur only for a short interval.

EXERCISE

1. Discuss the construction, principle and working of a valve type arrester.

2. What is a surge absorber ? Write a short note on Ferranti surge absorber.

3. Write short notes on the following surge diverters :
   (i)   Rod gap diverter
   (ii) Horn gap diverter
   (iii) Expulsion type diverter
   (iv) Multigap diverter

4. What are volt-time curves ? What is their significance in power system studies ?

5. What are BILS ? Explain their significance in power system studies.

6. Compare the relative performances of the following: (i) Rod gap; (ii) Expulsion gap; and (iii) Valve type L.A.

7. What do you mean by grounding or earthing? Explain it with an example.

8. Describe ungrounded or isolated neutral system. What are its disadvantages?

9. What is neutral grounding?

10. What are the advantages of neutral grounding?

11. What is solid grounding? What are its advantages?

12. What are the disadvantages of solid grounding?

13. What is resistance grounding? What are its advantages and disadvantages?

14. Describe Arc suppression coil grounding.

15. What is resonant grounding?

MODEL QUESTION PAPERS FOR
End-Semester Examination

Paper- I

Time : 3 Hrs. **Marks : 60**

Instructions to the candidates :

1. Each Question carries 12 Marks.

2. Attempt any five questions from the following.

3. Illustrate your answers with neat sketches, diagram etc., wherever necessary.

4. If some part or parameter is noticed to be missing, you may appropriately assume it and should mention it clearly.

1. **(a)** Explain with the help of neat diagram the construction and working of : **[6]**

   (i) Non-directional induction type overcurrent relay

   (ii) Induction type directional power relay

   (iii) Describe the construction and principle of operation of an induction type directional overcurrent relay.

   (b) Write a detailed note on differential relays. **[6]**

2. **(a)** Explain the difference between bulk oil circuit breakers and low-oil circuit breakers. **[6]**

   (b) In 132 kV transmission system, the phase to ground capacitance is 0.01 µF. . The inductance being 6 H. Calculate the voltage appearing across the pole of a circuit breaker if a magnetizing current of 10 A is interrupted. Find the value of resistance to be used across contact space to eliminate the striking voltage transient. **[6]**

3. **(a)** Discus briefly sampling theorem. **[6]**

   (b) Explain Surge Protection Circuits using digital protection **[6]**

4. **(a)** Do overhead systems need differential protection schemes than underground systems? **[6]**

   (b) Describe the following systems of bus-bar protection : **[6]**

   (i) Differential protection

   (ii) Fault-bus protection

5. **(a)** Describe the Merz-Price circulating current system for the protection of transformers. **[6]**

   (b) A generator is protected by restricted earth fault protection. The generator ratings are 13.2 kV, 10 MVA. The percentage of winding protected against phase to ground fault is 85%. The relay setting is such that it trips for 20% out of balance. Calculate the resistance to be added in the neutral to ground connection. **[6]**

6. **(a)** What is a surge absorber ? Write a short note on Ferranti surge absorber. **[6]**

   (b) Describe ungrounded or isolated neutral system. What are its disadvantages? **[6]**

MODEL QUESTION PAPERS FOR
End-Semester Examination

Paper- II

Time : 3 Hrs.   **Marks : 60**

Instructions to the candidates :

1. Each Question carries 12 Marks.
2. Attempt any five questions from the following.
3. Illustrate your answers with neat sketches, diagram etc., wherever necessary.
4. If some part or parameter is noticed to be missing, you may appropriately assume it and should mention it clearly.

1. **(a)** What is protective relay ? Explain its function in an electrical system. **[6]**

    (b) Write a brief note on relay timing. **[6]**

2. **(a)** Define and explain the following terms as applied to circuit breakers. **[6]**

    (i)   Arc voltage

    (ii) Restriking voltage

    (iii) Recovery voltage

    (b) A 50 Hz, 11 kV, 3-phase alternator with earthed neutral has a reactance of 5 ohms per phase and is connected to a bus-bar through a circuit breaker. The distributed capacitance upto circuit breaker between phase and neutral in 0.01 μF. Determine **[6]**

    (i)   Peak re-striking voltage across the contacts of the breaker

    (ii) Frequency of oscillations

    (iii) The average rate of rise of re-striking voltage upto the first peak.

3. **(a)** Explain numerical transformer differential protection with neat diagram. **[6]**

    (b) Write about Signal conditioning subsystem **[6]**

4. **(a)** Explain the Translay protection scheme for feeders. **[6]**

    (b) Discuss the time-graded overcurrent protection for **[6]**

    (i)   Radial feeders

    (ii) Parallel feeders

    (iii) Ring main system

5. **(a)** Explain with a neat diagram the application of Merz-Price circulating current principle for the protection of alternator. **[6]**

    (b) A star connected 3 phase, 12 MVA, 1 kV alternator has a phase reactance of 10%. It is protected by Merz-Price circulating current scheme which is set to operate for fault current not less than 200 A. Calculate the value of earthing resistance to be provided in order to ensure that only 15% of the alternator winding remains unprotected. **[6]**

6. **(a)** Write short notes on the following surge diverters : **[6]**

    (i)   Rod gap diverter

    (ii) Horn gap diverter

    (iii) Expulsion type diverter

    (iv) Multigap diverter

    (b) What are the advantages of neutral grounding? **[6]**

◈ ◈ ◈